A Minimalist Ontology of the Natural World

This book seeks to work out which commitments are minimally sufficient to obtain an ontology of the natural world that matches all of today's well-established physical theories. We propose an ontology of the natural world that is defined only by two axioms. *(1) There are distance relations that individuate simple objects—namely, matter points. (2) The matter points are permanent, with the distances between them changing.* Everything else comes in as a means to represent the change in the distance relations in a manner that is both as simple and as informative as possible. The book works this minimalist ontology out in philosophical as well as mathematical terms and shows how one can understand classical mechanics, quantum field theory and relativistic physics on the basis of this ontology. Along the way, we seek to achieve four subsidiary aims: (a) to make a case for a holistic individuation of the basic objects (ontic structural realism); (b) to work out a new version of Humeanism, dubbed Super-Humeanism, without natural properties; (c) to set out an ontology of quantum physics that is an alternative to quantum state realism and that avoids any ontological dualism of particles and fields; and (d) to vindicate a relationalist ontology based on point objects also in the domain of relativistic physics.

Michael Esfeld has been full professor of philosophy of science at the University of Lausanne since 2002. His last book with Routledge is *Conservative Reductionism* (with Christian Sachse) (2008).

Dirk-André Deckert is leader of the junior research group Interaction of Light and Matter in the Mathematical Institute of Ludwig Maximilians University Munich. He is the author of *Electromagnetic Absorber Theory – A Mathematical Study* (2010).

Routledge Studies in the Philosophy of Mathematics and Physics

Edited by:
Elaine Landry
University of California, Davis, USA
and
Dean Rickles
University of Sydney, Australia

For a full list of titles in this series, please visit www.routledge.com

A Minimalist Ontology of the Natural World

Michael Esfeld and Dirk-André Deckert
with Dustin Lazarovici, Andrea Oldofredi
and Antonio Vassallo

Routledge
Taylor & Francis Group

LONDON AND NEW YORK

First published 2018 by Routledge

2 Park Square, Milton Park, Abingdon, Oxfordshire OX14 4RN

52 Vanderbilt Avenue, New York, NY 10017

Routledge is an imprint of the Taylor & Francis Group, an informa business

First issued in paperback 2020

Library of Congress Cataloging-in-Publication Data
Names: Esfeld, Michael, author. | Deckert, Dirk-André, 1979– author. | Lazarovici, Dustin, 1985– author. | Oldofredi, Andrea, author. | Vassallo, Antonio, author.
Title: A minimalist ontology of the natural world / by Michael Esfeld and Dirk-André Deckert, with Dustin Lazarovici, Andrea Oldofredi, and Antonio Vassallo.
Description: 1 [edition]. | New York : Routledge, 2017. | Series: Routledge studies in the philosophy of mathematics and physics ; 3 | Includes bibliographical references and index.
Identifiers: LCCN 2017031890 | ISBN 9781138307308 (hardback : alk. paper)
Subjects: LCSH: Quantum theory—Philosophy. | Physics—Philosophy.
Classification: LCC QC174.12 .E84 2017 | DDC 530.12—dc23
LC record available at https://lccn.loc.gov/2017031890

ISBN: 978-1-138-30730-8 (hbk)
ISBN: 978-0-367-59412-1 (pbk)

Typeset in Sabon
by Apex CoVantage, LLC

Contents

1 Introduction

1.1 The aim of the book

Physics, being the study of nature (*physis* in Greek), and metaphysics, being the study of the most fundamental and general traits of being (cf. Aristotle, *Metaphysics* IV, 1003a21-22), can together be expected to answer the following three questions:

1. What is matter? What is space and time?
2. What are the laws of nature?
3. How does matter in space and time, being subject to certain laws, explain the observable phenomena?

The question of what is matter is connected with the question of what is space and time: one's view of matter has implications for one's view of space and time, and *vice versa*. The stance that one takes with respect to laws of nature, by contrast, is largely independent of the stance that one takes with respect to matter, space and time. The question about the laws has two aspects: what their content is and what their status of being is. Finally, the position that one endorses with respect to matter, space and time as well as laws is probed by the way in which one elaborates on how this position as a whole explains the observable phenomena.

These questions are philosophical as well as scientific. Even if it may seem too optimistic to hope that one day answers to these questions can be given in full completeness, the mere attempt of finding sensible answers and weighing their successfulness is already a fruitful enterprise in building a view of nature and its workings. This enterprise can best be characterized as *natural philosophy*, involving physics and philosophy in a seamless manner.

This enterprise is probably as old as mankind, and possible answers range, say, from quantum mechanics to religion. This is due to the fact that it is not even clear in which sense we would be satisfied with a potential answer. Since human thought is tied to concepts, one of the first questions should be which concepts should we employ in our formulation? Are "particles, fields, strings, genes, trees, souls, devils, angels, and gods" good choices?

Obviously, the story we will attempt to tell in finding answers to the aforementioned questions will depend heavily on this choice, and some of the chosen words should better relate to some of the natural things that exist independently of our thoughts and language; otherwise, the aforementioned enterprise will be futile. On the one hand, the larger the vocabulary that has to be taken as basic, the shorter the potential stories, but also the weaker is their explanatory power as the many concepts in the vocabulary cannot be scrutinized. On the other hand, the smaller the basic vocabulary, the more can be scrutinized, but also the longer will the stories be that we have to tell. Consider how Jackson (1994) describes the enterprise of metaphysics:

> Metaphysics, we said, is about what there is and what it is like. But of course it is concerned not with any old shopping list of what there is and what it is like. Metaphysicians seek a comprehensive account of some subject matter—the mind, the semantic, or, most ambitiously, everything—in terms of a limited number of more or less basic notions. In doing this they are following the good example of physicists. The methodology is not that of letting a thousand flowers bloom but rather that of making do with as meagre a diet as possible. . . . Because the ingredients *are* limited, some putative features of the world are not going to appear explicitly in the story. The question then will be whether they, nevertheless, figure implicitly in the story. Serious metaphysics is simultaneously discriminatory and putatively complete, and the combination of these two facts means that there is bound to be a whole range of putative features of our world up for either elimination or location.
>
> (Jackson (1994), p. 25)

The same goes for the scientific enterprise in general. The sciences seek a good balance between the complexity of the vocabulary and the length of their explanations that should of course depend on the natural phenomena under study. For instance, while molecules may be well described by using the notion of atoms, it would be a tiresome endeavour to explain the functions of a cell in terms of only the atomistic vocabulary. Such switches in the basic concepts that are employed are quite common, even in mathematics. However, they often make it difficult to relate one well-established theory to another one. Consider, to mention just one example, how difficult it is to specify how and when quantum mechanics can be well approximated by Newtonian mechanics.

Nevertheless, all these theories have a common goal—namely, to describe what there is. As it would be outrageous to think that what there is depends on our theories about the world, there must be a sense in which all these vocabularies have a, in some sense minimal, common set relating to the things that exist. One convincing example that such a common set can be found comes from statistical mechanics, which is able to relate the concepts "temperature, pressure, volume and entropy" used in thermodynamics to the concept "particle motion" used in Newtonian mechanics. A common set of concepts must contain good candidates for building an ontology of the natural world. Although we humans may never be able to fully infer what there really is, we can at least ask the following question. What is a minimal set of entities that form an ontology that matches today's well-established physical theories? In answering this question, we will carefully distinguish within the vocabulary used in these physical theories between, on the one hand, the concepts that relate to what there is in the sense of this minimal set of entities and, on the other hand, the concepts that make up what we call the dynamical structure of a physical theory, providing an economic means of telling the necessary scientific stories.

In this sense, using parsimony as the guide for ontology, the aim of this book is to develop a minimalist answer to the aforementioned questions: *we seek to work out which commitments are minimally sufficient to obtain an ontology of the natural world that is empirically adequate.* Generally speaking, the reason for employing parsimony as the guide for ontology is that for any candidate entity stemming from science—or common sense, or intuitions—we need an argument for why one should endorse an ontological commitment to that entity. Its being part of what is minimally sufficient to obtain an ontology of the natural world that is empirically adequate is the best argument for an ontological commitment. It is an illusion to think that by abandoning parsimony and enriching the ontology, one achieves explanations that are deeper than those that a parsimonious ontology can yield; one thereby runs only into artificial problems and impasses, as we shall show in this book.

In this vein, we start from the idea that given a plurality of objects, there has to be a certain type of relations in virtue of which these objects make up a world. The minimalist hypothesis then is that these relations also individuate the objects, thus paving the way for the claim that there is nothing more to these objects than standing in these relations. The objects thus are simple, having no parts or any other internal structure. When it comes to the natural world, relations providing for extension—namely, distances—are the first and foremost candidate for the type of relations that fulfills this task. Distances connect unextended and thus point-sized objects. If they individuate these objects, they provide for variation within a configuration of point-sized objects, with each of these objects being distinct from all the other ones by at least one distance relation that it bears to another object. In virtue of standing in distance relations, these objects then are matter points (recall the sparse Cartesian conception of the natural world as *res extensa*). In order to achieve empirical adequacy, we furthermore have to stipulate that these relations change. We thus propose an ontology of the natural world that is defined by the following two axioms, and only by these two axioms:

(1) *There are distance relations that individuate simple objects—namely, matter points.*
(2) *The matter points are permanent, with the distances between them changing.*

We submit that these two axioms prescribe the diet that is as meagre as possible in accounting for the natural world, to come back to the citation from Jackson earlier. Everything else then comes in as a means to represent the change in the distance relations that actually occurs in a manner that is both as simple and as informative as possible.

We thus take up atomism and seek to develop it into a minimalist ontology of the natural world. Atomism is the oldest and most influential tradition in natural philosophy, going back to the pre-Socratic philosophers Leucippus and Democritus. The latter is reported as maintaining that

> substances infinite in number and indestructible, and moreover without action or affection, travel scattered about in the void. When they encounter each other, collide, or become entangled, collections of them appear as water or fire, plant or man.
> (fragment Diels-Kranz 68 A57, quoted from Graham (2010), p. 537)

In a similar vein, Newton writes at the end of the *Opticks*,

> It seems probable to me, that God in the Beginning form'd Matter in solid, massy, hard, impenetrable, moveable Particles . . . the Changes of corporeal Things are to be placed only in the various Separations and new Associations and motions of these permanent Particles.
> (Newton (1952), question 31, p. 400)

The attractiveness of atomism is evident from these quotations: on the one hand, it is a proposal for a fundamental ontology that is most parsimonious and most general. On the other hand, it offers a clear and simple explanation of the realm of our experience. Macroscopic objects are composed of indivisible particles. All the differences between the macroscopic objects—at a time as well as in time—are accounted for in terms of the spatial configuration of these particles and its change, which is subject to certain laws. That is why Feynman famously writes at the beginning of the *Feynman lectures on physics*,

> If, in some cataclysm, all of scientific knowledge were to be destroyed, and only one sentence passed on to the next generations of creatures, what statement would contain the most information in the fewest

words? I believe it is the *atomic hypothesis* (or the atomic *fact*, or whatever you wish to call it) that *all things are made of atoms—little particles that move around in perpetual motion, attracting each other when they are a little distance apart, but repelling upon being squeezed into one another.* In that one sentence, you will see, there is an enormous amount of information about the world, if just a little imagination and thinking are applied.

(Feynman et al. (1963), ch. 1–2)

Whereas atomism is a purely philosophical proposal in Leucippus and Democritus, it is turned into a precise physical theory by Newton. Accordingly, classical mechanics—and classical statistical physics—are usually seen as the greatest triumph of atomism. Nonetheless, atomism loses nothing of its attractiveness when it comes to quantum physics. In the first place, also in this domain, all experimental evidence is evidence of discrete objects (i.e. particles)—from dots on a display to traces in a cloud chamber. Entities that are not particles—such as waves or fields—come in as figuring in the explanation of the behaviour of the particles, but they are not themselves part of the experimental evidence: an electric field is probed by the motion of a test charge subject to it (such as the electron in the wire); the double slit experiment is made apparent by sufficiently many particles hitting on a screen, etc.

Moreover, in quantum as in classical physics, there are good arguments to maintain that we need an account of macroscopic objects—such as, for instance, a cat or an apparatus with a pointer that points in a certain direction—in terms of matter being arranged in a certain manner in physical space. In order to achieve such an account, one cannot only endorse the quantum state, which is defined on a very high-dimensional mathematical space (namely, the configuration space of the universe), but one has to conceive that state as being the state of matter arranged in three-dimensional space or four-dimensional space-time: we take the arguments to this effect going back to Bell (2004, ch. 7) and elaborated on notably by Maudlin (2010, 2015) to be convincing. In brief, according to these arguments, it is not sufficient that one can find something in the quantum state of the universe that functionally corresponds to cat-like or pointer-like behaviour; for there to be a cat, or a pointer, there have to be basic objects that compose a cat, or a pointer, if they are arranged in the right manner in physical space such that the evolution of such a configuration of basic objects then amounts to the motion of a cat, or a pointer, in space.

In other words, then, what has become known as a *primitive ontology* of matter distributed in three-dimensional space or four-dimensional space-time is a necessary condition to avoid the famous measurement problem of quantum physics.[1] To turn that necessary condition into a sufficient one, one has to formulate a dynamic for the primitive ontology that excludes superpositions of matter in space so that there always is a definite configuration of matter and Schrödinger's cat paradox, among others, is avoided, but that dynamic has to include entanglement to account for the non-local correlations that are manifest, for instance, in the Einstein Podolsky Rosen (EPR) experiment (for recent elaborations, see, notably, Allori et al. (2008), Belot (2012) and Esfeld et al. (2014)). This reasoning applies not only to non-relativistic quantum mechanics but also to quantum field theory (QFT)—as well as a future theory of quantum gravity—since any quantum theory is plagued by the measurement problem (cf. Barrett (2014)).

Its success notwithstanding, atomism faces three major problems. Our transformation of atomism into a truly minimalist ontology of the natural world seeks to address these problems.

(1) In the first place, Democritus as well as Newton set out atomism in terms of a dualism of matter on the one hand and space and time on the other: matter is conceived as being inserted in an absolute background space and as evolving in an absolute background time. However, the justification of absolute space and time is debatable, and their ontological status remains unclear in classical atomism. The commitment to absolute space and time implies in any case a commitment to a surplus structure, because absolute space and time reach, in any case, far beyond the actual configuration of matter, with its being doubtful whether one obtains a gain in explanation through that commitment.

Since Leibniz, *relationalism* about space and time is put forward to avoid that dualism. We follow this tradition. We set out an ontology of the natural world in relationalist terms, being committed only to distance relations among the atoms and deriving time from the change in these relations as the order of that change. We show how such an ontology can match both classical and quantum mechanics and how it remains a viable option in relativistic physics.

(2) Nonetheless, even if the dualism of matter on the one hand and space and time on the other is removed, the question of what characterizes the atoms as material substances remains. Democritus and Newton conceive them as being equipped with a few basic intrinsic properties—that is, properties that belong to each atom taken individually, independently of all the other ones, thus making up an intrinsic essence of each atom. The paradigmatic example is mass in Newtonian mechanics. However, also in Newtonian mechanics, both inertial and gravitational mass are introduced through their dynamical role—namely, as a dynamical parameter that couples the motions of the particles to one another, as was pointed out by Mach (1919, p. 241) among others. The same goes for charge, energy, etc. When it comes to quantum mechanics, despite first appearances, properties such as mass and charge cannot be conceived as intrinsic properties of the particles, but are situated on the level of their quantum state as represented by the wave function. In sum, as soon as atomism is worked out as a precise physical theory, it turns out that anything that one might regard as constituting an intrinsic essence of the atoms is in fact a dynamical

parameter, expressing a dynamical relation that couples the motions of the atoms to one another. Hence, the question is what is the essence of the atoms qua material entities?

We bring in *ontic structural realism* to answer this question: instead of having an intrinsic essence, the atoms have a structural one. Standing in distance relations is their essence. Hence, although we propose an ontology of atomism, we draw on *holism* to work that ontology out: the atoms are holistically individuated in terms of the distances among them. We conceive the distance relations as establishing the order of what coexists, thereby taking up Leibniz's relationalist definition of space: these relations are able to distinguish the objects, thereby satisfying the principle of the identity of indiscernibles. There thus is a configuration of objects that is constituted by distance relations: by individuating the atoms, the distance relations provide for *variation* within a given configuration of matter.

Over and above variation making up for a configuration of objects, there is *change*, which hence is change in the relations that constitute the configuration—that is, the distances. We follow Leibniz in conceiving time as the order of that change, with that order being unique and having a direction. Mass, charge, energy, spin, wave function, etc., then, are dynamical parameters that a physical theory introduces in order to obtain a law that describes that change in a simple and informative manner. These parameters sort the atoms into different particle species on the basis of salient patterns in their relative motion. Consequently, the atoms are not intrinsically protons, electrons, neutrons, etc., but are so described because their motion exhibits certain contingent regularities. In a nutshell, some atoms do not move electronwise because they are electrons, but they can be classified as electrons because they move electronwise.

Indeed, *there is no need to admit physical properties at all*. Relations do all the work. It is a misconception to set out ontic structural realism as a stance that is directed against object-oriented metaphysics (cf. Ladyman and Ross (2007) and French (2014)). Ontic structural realism is opposed to the property-oriented metaphysics that has dominated philosophy from Aristotle to today's analytic metaphysics. Of course, if there are relations, there are objects that stand in the relations, but standing in the relations is all there is to these objects—the relations are their essence (cf. the moderate ontic structural realism set out in Esfeld (2004), Esfeld and Lam (2008, 2011)).

In order to understand what physics says about the natural world, one should not be misled by the subject-predicate form of ordinary language to buying into an ontology of substances that are characterized by intrinsic properties. The simple and elegant story that physics tells from its beginnings to this day is one of discrete objects—call them "matter points"—standing in distance relations and dynamical parameters capturing the change in these relations. The dynamical parameters—that is, our attempts to conceive a dynamical structure that describes this change in a simple, elegant and informative manner—vary from one theory to another. By contrast, the commitment to simple, discrete objects standing in distance relations whose evolution these dynamical parameters seek to track remains constant, from Leucippus and Democritus via Newton to today's quantum theories.

That is to say, *the argument from theory change and pessimistic meta-induction against scientific realism does not apply to our minimalist ontology and neither does the argument from underdetermination.* What is underdetermined by the given evidence is the best formulation of a dynamical structure, but not the ontology of relative particle positions and motion—that is, matter points individuated by distance relations and change in these relations. Consequently, we draw within any given physical theory a distinction between the primitive variables that directly refer to what there is in the world (i.e. the primitive ontology)—namely, relative particle positions and their change—and the laws that describe the evolution of the primitive variables (i.e. the dynamical structure of a physical theory), with all the other variables being nomological in the sense that they fall on the side of the laws instead of directly referring to something that there is in the world over and above the primitive variables.

Hence, in a nutshell, there are distance relations individuating the matter points and thereby constituting a configuration of them, and there is change in these relations. *That is all.* In terms of Humeanism, the distance relations among the matter points and their change throughout the entire history of the universe are the Humean mosaic, and everything else in the natural world supervenes on that mosaic in the sense that it comes in as a means to describe that change in a manner that is both as simple and as informative as possible. *The argument for this sparse ontology is its simplicity together with its empirical adequacy: less won't do for an ontology of the natural world; bringing in more creates new drawbacks instead of providing additional explanatory value.* This sparse ontology hence amounts to a radical ontological reductionism: everything in the natural world reduces to distance relations among matter points and the change in these relations, in the sense that these relations and their change make true all the true propositions about the natural world. (By the natural world, we mean the physical, spatially extended world. We have no intention here to apply this reductionism to the mind, consciousness and normativity).

(3) However, an ontology of distance relations among point particles (matter points) and their change only may appear as hopelessly old-fashioned and therefore obviously wrong. Newtonian mechanics is not all of classical physics. There also is classical electrodynamics, apparently refuting atomism as a complete ontology of the natural world by replacing the commitment to atoms only with a commitment to both particles and fields being the stuff out of which the world is made. But this dualism runs into an impasse: as Feynman stresses in his Nobel lecture, the field

spreads out to infinity, being defined everywhere in space-time, thus also in regions where there will never be any particles whose motion it influences (Feynman (1966), pp. 699–700). What then is the field? Is it some sort of stuff filling all of space-time? Is it a property of space-time points, albeit not a geometrical one? In the former case, we are committed to an extravagant ontology with there being stuff in addition to the particles everywhere, although all the experimental evidence that we have is one of particle stuff only, whose relative distances change. In the latter case, we are back to the commitment to an absolute background space as the carrier of the field properties (see Field (1985), pp. 40–42), which again exist everywhere in space-time—that is, also in regions where they never manifest themselves by influencing the motion of particles.

Apart from these questions there are physical and mathematical issues that have not been solved till today. The influence that a particle exerts on the electromagnetic field reacts back on the particle, leading to an infinite force at the position of the particle and, hence, ill-defined equations of motion. These difficulties can as yet only be dealt with in a perturbative sense by means of a so-called renormalization of ill-defined corrections to the non-interacting theory. These problems are inherited by QFT and string theory (to date, there is no non-perturbative QFT describing a non-trivial relativistic interaction in 3 + 1 dimensions due to this problem). The philosophical problem of the ontological status of fields becomes even worse in QFT: the quantum state, represented by the wave function, can no longer be considered as a field in four-dimensional space-time as in the case of quantum mechanics of one particle, but takes as arguments elements of the space of configurations of fields.

The dualism of a quantum state (wave function) that is a field in configuration space and particles in ordinary space is even more unconvincing than the classical dualism of particles and fields since it is mysterious how objects in different spaces could interact. This implausible dualism is since Everett (1957) the motivation for recognizing only the quantum state (see most recently Albert (2015), chs. 6 and 7, for this motivation). However, as the ongoing discussion shows, it is by no means obvious how one can account for our experience and make sense of the quantum mechanical probability calculus on the basis of recognizing only the quantum state.[2] Nonetheless, note that the parsimonious ontology advocated in this book has a much wider scope than what is known as the primitive ontology approach to quantum mechanics. Notably, it is not tied to three-dimensional space; the distance relations defining this ontology are not wedded to a particular geometry. Hence, from the perspective of this ontology, the main objection to a quantum ontology of only a quantum state in configuration space is not that this approach lacks a primitive ontology in terms of objects that are localized in three-dimensional space. The main objection does not concern the dimension of the underlying space, but the uneconomical dualism of a substantival space and material entities (such as a wave function field) defined on that space.

The alternative move to a quantum ontology in terms of the quantum state only is to draw on the other main ontology of quantum physics that has also been pursued since the 1950s—namely, Bohm's theory, which lays stress on particles in ordinary space and thereby avoids from the start any form of a quantum measurement problem: there always is a definite spatial configuration of matter (no superpositions), composing the macroscopic objects with which we are familiar. More generally speaking, given that all the evidence in classical as well as quantum physics is evidence of relative particle positions and motion, the most straightforward explanation is the one that takes this evidence literally in terms of a particle ontology—although, of course, particles *qua* matter points individuated by distance relations are theoretical entities. They are introduced because they constitute the most parsimonious ontology that enables the formulation of a dynamics such that the way in which the particles move then explains the available particle evidence in the sense that this ontology is, taking everything into account, overall more coherent than its rival, richer ontologies. In particular, we show how an ontology of permanent particles that move on continuous trajectories according to a deterministic law is also able to explain the experimental evidence in QFT, including notably the appearance of particle creation and annihilation events.

However, as this ontology is set out in Bohm (1952a) (and earlier in de Broglie (1928)), it directly runs into the unconvincing dualism of particles in ordinary space and a wave function in configuration space. Basing ourselves on the dominant contemporary formulation of this theory known as Bohmian mechanics by Dürr et al. (2013b) and their argument to regard the wave function as nomological, we tackle this problem by deleting the commitment to the wave function as a physical entity in addition to and on a par with the particles: the wave function or quantum state is nothing but a dynamical parameter introduced in the theory to capture the evolution of the particles' positions—that is, the change in their distances. In a nutshell, instead of overcoming the implausible dualism of matter in ordinary space and quantum state in configuration space by recognizing only the quantum state, we submit that an ontology of matter in physical space only, with that space being nothing but the distance relations among matter points and the change in these relations, is sufficient and indeed the best available option to account for the experimental evidence in quantum as well as classical physics.

The central aim of this book, accordingly, is to elaborate on an ontology of the natural world that is most parsimonious while being empirically adequate: a commitment to less than distance relations among matter points that are individuated by these very relations and change in these relations would be insufficient. A commitment to more than that leads to trouble instead of yielding deeper explanations. Along the way, we seek to achieve four subsidiary aims: (a) to make a case for a holistic individuation of the basic objects, combining ontic structural realism with relationalism about space; (b) to work out a new version of Humeanism, dubbed Super-Humeanism, that covers the whole of physics by doing without natural properties, taking the Humean mosaic to consist only in basic relations among point objects (i.e. distance relations among matter points) and the change in these relations; (c) to set out an ontology of quantum physics, including QFT, that is a complete alternative to Everett-style quantum state realism and that in general avoids any dualism of particles and fields by relegating the latter (including the quantum state) to the dynamical structure of a physical theory whose only function is to capture the change in the distance relations among the particles in ordinary space; and (d) to vindicate a relationalist ontology based on distances also in the domain of relativistic physics.

1.2 The content of the book and its methodology

Apart from the introduction, the book has four chapters: in the second chapter, we develop the minimalist ontology of matter points and their dynamics in general philosophical as well as mathematical terms, independently of any particular physical theory. We thereby set out a new argument for relationalism about space and time, ontic structural realism about the basic objects and Humeanism about the physical laws and the dynamical parameters figuring in them. In Chapter 3, we show how this ontology fits both classical mechanics of gravitation and quantum mechanics, as long as the latter includes a primitive ontology of matter distributed in ordinary space, as in Bohmian mechanics. We point out the similarities between these two theories as well as their differences and explain how in both cases one gets from an ontology and a dynamics defined for the universe as a whole to a description of subsystems of the universe and a dynamics for them.

Chapters 4 and 5 then expand on this ontology with respect to today's most advanced, established theoretical physics. In Chapter 4, we go into QFT: in terms of the Dirac sea model for electrons, we show how an ontology of persisting particles that move according to a deterministic law can explain the characteristic phenomena of QFT—in particular the appearance of particle creation and annihilation—and ground its empirical predictions. In Chapter 5, we consider relativistic physics, sketching out a version of relationalism about space and time combined with Super-Humeanism also for general relativistic physics on the basis of an ontology of matter qua matter points that are individuated by the distance relations in which they stand and characterized by the change in these relations.

Our methodology is the one of *natural philosophy* or *naturalized metaphysics*, treating physics and metaphysics as going together in an inseparable, seamless manner. We adopt an empiricist attitude in insisting on the fact that all the experimental evidence consists in relative particle positions and motion, working out a minimally sufficient ontology that accounts for this evidence and arguing that any additional ontological commitment comes with new drawbacks instead of providing a deeper explanation. However, this book is opposed to a neo-positivist metaphysics as set out notably in Ladyman and Ross (2007), in the sense of the idea that there is a one-way road leading from physics to metaphysics, suggesting that one obtains metaphysical claims from the mathematical structures of physical theories and that only metaphysical claims thus established are valid.[3] Consequently, we reject Quine's criterion of ontology in terms of, in brief, formulating a theory in first-order logic and being committed to all the entities that the theory thus formulated quantifies over (see Quine (1948)).

In general, on the one hand, this book is directed against the idea that any metaphysical claims—be they positive, seeking to tell us what there is, be they negative, seeking to rule out certain ontological options—can be established directly on the basis of the mathematical structure of physical theories. One cannot infer ontology directly from the mathematical structure of a physical theory, as, for instance, the futile attempts of making ontological claims about trajectories, individuals, probabilities, etc., as well as setting up various no-go theorems on the basis of the operator formalism of textbook quantum mechanics show (see Laudisa (2014) for a recent criticism). But also a theory that goes beyond textbook quantum mechanics, such as Bohmian mechanics, does not simply wear the ontology on its sleeves, as the ongoing debate about the status of the wave function in this theory illustrates. On the other hand, the *a priori*, armchair metaphysics based on conceptual analysis that is widespread in today's analytic philosophy cannot on its own produce knowledge of the world, for it lacks the means to cast ontological claims in a precise physical framework that is able to predict and explain empirical phenomena. That is why we promote natural philosophy—metaphysics becoming physics, or physics being done on the basis of first ontological principles.

We advocate drawing a distinction between what simply exists according to a given theory—that is, the *primitive ontology*—and what is introduced in the theory through its functional (or dynamical) role for what there simply is—that is, the *dynamical structure* of the theory, as expressed in its mathematical formalism. Thus, a wave function in quantum mechanics as well as parameters such as mass and charge in classical mechanics presuppose a configuration of objects to which they are ascribed because they are introduced in terms of their functional or dynamical role for the evolution of these objects; by contrast, these objects—that is, the point particles—do not have any further function in the theory. They are simply there.

In a second step, we move from primitive ontology within a theory to ontology *tout court*. Natural philosophy seeks for an ontology of the natural world that is not relative to particular physical theories. To our mind, it is inappropriate to speak of the ontology of this or that physical theory. Ontology is about what there is. It goes without saying that our access to what there is comes through the representations that we conceive in terms of physical theories. But this does not imply that ontology is relative to particular theories, because it is a mistake to seek to read ontology off from the mathematical formalisms of physical theories. *The measure for ontology is simplicity (or parsimony) together with empirical adequacy.* Of course, parsimony is an *a priori* criterion of armchair metaphysics (cf. Jackson (1994), p. 25). However, there is nothing wrong with using that criterion as long as one can show in detail how it is possible to get from the ontology formulated on that basis to the established physical theories and, moreover, thereby solve the problems of interpretation of these theories (such as the quantum measurement problem). The argument for parsimony then is that admitting more to the ontology than what is minimally sufficient for the purpose of empirical adequacy does not amount to a gain in explanation. It only leads to drawbacks (such as, e.g., the ones pointed out by Leibnizians against absolute space), or creates artificial problems (such as, e.g., the questions of how a particle can reach out to other particles and accelerate them in virtue of properties that are intrinsic to it, or how a wave function, being defined as a field on configuration space, can push particles around in physical space).

Obviously, the measure for representation also is simplicity together with empirical adequacy. *However, simplicity in ontology and simplicity in representation pull in opposite directions.* That is why the neo-positivist trend of using the mathematical structure of a physical theory as the guideline for ontology is wrong-headed. Employing only the concepts that describe what there is on the simplest ontology (i.e. matter points individuated by distance relations), the description of the evolution of the configuration of matter would not be simple at all, since one could not do much better than listing all the change that actually occurs. Reading one's ontological commitments off from the simplest description—such as, e.g., Newtonian mechanics—the ontology would not be simple at all: it would in this case be committed to absolute space and time, to momenta, gravitational masses, forces, etc.

To put this crucial issue differently, from the epistemological perspective, in metaphysics as well as in physics or mathematics, we have to start from a few basic notions taken as primitive. We can elucidate what these notions mean—as we shall do with the notion of distances individuating point particles in section 2.1—but we cannot trace them back to other notions. These notions, then, make up our proposal for the primitive ontology: what there simply is. It is in principle possible to do the whole of physics with just the notions of distances individuating point particles and change of these distances, but this would be uneconomical: no simple dynamical equation capturing that change could be thus achieved. The reason is that we can define a configuration of matter and its change with these basic notions, but there is nothing in a configuration of matter given in terms of the relative distances between the matter points that provides information about how the change of that configuration occurs. It is therefore reasonable to introduce further notions that provide such information by being formulated in terms of their functional role for the change in the configuration of matter so that the representation of change becomes more simple without losing information. These notions, which make up the dynamical structure of a physical theory, vary as we acquire more data and make progress in constructing a system that strikes as good as we can achieve a balance between being simple and being informative about the change in the world.

However, it is a misunderstanding to think that by introducing further notions one subscribes to ontological commitments that go beyond the ones given in terms of the basic notions. This, again, is an instance of the neo-positivist fallacy of inferring ontology from the mathematical structure of a theory. Here is the reason Humeanism enters into this book—namely, as a strategy to maintain scientific realism without building ontological commitments on the representational means that physical theories employ. If we take the basic notions to define what is known as the Humean mosaic, we can conceive all the further notions that we need to achieve a simple theory—namely, the geometrical and the dynamical ones—as being the means to obtain a representation of that mosaic that is both simple and informative. We thereby stay neutral with respect to the issue of whether or not there is something modal in the world—a *logos* in the *cosmos* so to speak. We only claim that it is misguided to conceive the varying means that we set up to conceptualize the change that we perceive in a simple and informative manner as revealing that *logos*. Doing so only leads to artificial problems, such as the ones involved in the surplus structure that comes with the commitment to absolute space and time, or the ones showing up in the question of how a wave function can push particles around in physical space.

1.3   Acknowledgements

The work on writing this book began when Michael Esfeld spent the academic year 2014–15 in Munich, having received the research award of the Alexander von Humboldt Foundation. Michael Esfeld is grateful to Stephan Hartmann and the Munich Center for Mathematical Philosophy for their hospitality, as well as to the Humboldt Foundation whose award and continuing support made this work possible. The overall framework of this book was conceived by Michael Esfeld and Dirk-André Deckert, with Michael Esfeld taking the lead for the philosophical argument and Dirk-André Deckert for the physical one. Chapter 3, section 2, on identity-based Bohmian mechanics goes back to work by Dustin

Lazarovici, as do Chapter 3, section 4, and Chapter 5, section 2. Andrea Oldofredi joined in for collaboration on Chapter 4 and Chapter 3, section 4. Antonio Vassallo was the expert for setting out the relationalism for both classical mechanics and general relativistic physics. These collaborators are not to be held responsible for the overall framework of this book. Dirk-André Deckert's work was funded by the junior research group grant *Interaction between Light and Matter* of the Elite Network of Bavaria. Andrea Oldofredi's and Antonio Vassallo's contributions were supported by the Swiss National Science Foundation, grant no. 105212-149650. Dustin Lazarovici acknowledges funding from the Cogito foundation (grant no. 15-106-R) as well as the Alexander von Humboldt Foundation (Feodor Lynen research fellowship).

The book draws on material from the following papers: "Quantum Humeanism, or: Physicalism without Properties", by Michael Esfeld, published in the *Philosophical Quarterly* 64 (2014), pp. 453–470; "The Ontology of Bohmian Mechanics", co-authored by Michael Esfeld, Dustin Lazarovici, Mario Hubert and Detlef Dürr, published in the *British Journal for the Philosophy of Science* 65 (2014), pp. 773–796; "From the Universe to Subsystems: Why Quantum Mechanics Appears More Stochastic than Classical Mechanics", co-authored by Andrea Oldofredi, Dustin Lazarovici, Dirk-André Deckert and Michael Esfeld, published in *Fluctuations and Noise Letters* 15 (2016), Special issue Quantum and classical frontiers of noise, pp. 164002: 1–16; "The Physics and Metaphysics of Primitive Stuff", co-authored by Michael Esfeld, Dustin Lazarovici, Vincent Lam and Mario Hubert, published in the *British Journal for the Philosophy of Science* 68 (2017), pp. 133–161; "What Is Matter? The Fundamental Ontology of Atomism and Structural Realism", co-authored by Michael Esfeld, Dirk-André Deckert and Andrea Oldofredi, forthcoming in Anna Ijjas and Barry Loewer (eds.): *A Guide to the Philosophy of Cosmology*, Oxford University Press; "Relationalism about Mechanics Based on a Minimalist Ontology of Matter", co-authored by Antonio Vassallo, Dirk-André Deckert and Michael Esfeld, published in the *European Journal for Philosophy of Science* 7 (2017), pp. 299–318; "Leibnizian Relationalism for General Relativistic Physics", co-authored by Antonio Vassallo and Michael Esfeld, published in *Studies in History and Philosophy of Modern Physics* 55 (2016), pp. 101–107; "A Persistent Particle Ontology for QFT in Terms of the Dirac Sea", co-authored by Dirk-André Deckert, Michael Esfeld and Andrea Oldofredi, forthcoming in the *British Journal for the Philosophy of Science*, DOI 10.1093/bjps/axx018; "A Proposal for a Minimalist Ontology", by Michael Esfeld, forthcoming in *Synthese*, DOI: 10.1007/s11229-017-1426-8. We are grateful to the publishers of these papers to leave us the right to use parts of the material of these papers as a basis for this book.

We would like to thank the participants of the summer schools on the philosophy of physics in Saig (Black Forest, Germany) in July 2014, July 2015, July 2016 and July 2017 for discussions, in particular, David Albert,

Jeff Barrett, Jeremy Butterfield, Detlef Dürr, Matthias Egg, Tiziano Ferrando, Shelly Goldstein, Mario Hubert, Vincent Lam, Anna Marmodoro, Tim Maudlin, Ward Struyve and Nino Zanghì, as well as Peter Pickl, Christian Sachse, Barry Loewer, Harvey Brown, Oliver Pooley and Hans Christian Öttinger. Last, but not least, we are grateful to three anonymous referees of the draft of the book for their helpful comments and to Andrew Weckenmann for the professional handling of the book manuscript.

Notes

1. The term "primitive ontology" goes back to Dürr et al. (2013b, ch. 2), originally published 1992. A forerunner of this notion can be found in Mundy (1989, p. 46). Cf. also Bell's notion of "local beables" in Bell (2004, ch. 7), originally published 1975.
2. See Wallace (2012) as well as Wilson (2013a), Albert (2015), chs. 6 and 7, and Ney (2015) for proposals in this sense. See, notably, Maudlin (2010, 2014, 2015) as well as Dawid and Thébault (2014) and Dizadji-Bahmani (2015) for criticism.
3. See Ney (2012) for a discussion of neo-positivist metaphysics as well as a plea for a moderate version of this project, in contrast to the one put forward by Ladyman and Ross (2007). See, furthermore, Callender (2011), Morganti (2013) and Guay and Pradeu (2017) for more balanced views of naturalized metaphysics, as well as the papers in Ross et al. (2013), in particular, Chakravartty (2013), Humphreys (2013) and Wilson (2013b).

2 Matter points and their dynamics

2.1 Matter points individuated through distance relations

Let us come back to the quotations from Democritus and Newton in the introduction. As is evident from these quotations, they both consider atoms as the smallest, indivisible objects that compose all the other objects and whose spatial configuration explains the observable features of the macroscopic objects with which we are familiar and their change. But what then are the atoms? Both Democritus and Newton are usually taken to conceive them as being characterized by a few basic intrinsic properties—that is, properties that each atom has in itself, independent of its relationship to other atoms (see Langton and Lewis (1998) on intrinsic properties, and Hoffmann-Kolss (2010), part 1, for an extensive discussion). The paradigmatic example of such a property in classical physics is mass in Newtonian mechanics, with a value of mass attributed to each particle taken individually.

However, it is doubtful whether this view holds up to scrutiny. Also in Newtonian mechanics, both inertial and gravitational mass are introduced through their dynamical role—namely, as dynamical parameters that couple the motions of the particles to one another—as was pointed out by Mach (1919, p. 241) among others: in short, mass tells us something about how the particles move. Hence, mass is a parameter that expresses a dynamical relation between the atoms. The same goes for charge. Consequently, in expressing a dynamical relation, parameters such as mass and charge *presuppose* objects given in terms of their spatial location to which they are applied.

This argument against intrinsic properties is strengthened by taking quantum mechanics into consideration. Over and above mass and charge being parameters that express a dynamical relation between the particles, quantum mechanics contradicts the view that these parameters are situated at the particle locations and belong to the particles taken individually. In the first place, the wave function is the central dynamical parameter of quantum mechanics. Due to its being entangled, it is in general not possible to attribute a wave function or quantum state to the particles taken individually, but only to their configuration as a whole. Moreover, even in cases in which a many-particle wave function is separable, quantities like mass and charge are not located at the position of the particles, but spread over the support of the wave function, as far as they figure in particle interactions. Experimental considerations involving interference phenomena—for instance, in the context of the Aharonov-Bohm effect and of certain interferometry experiments—show, in brief, that mass and charge are effective at all the possible particle locations that the quantum state admits (see, e.g., Brown et al. 1995, 1996 and references therein; cf. most recently Pylkkänen et al. 2015). Consequently, also dynamical parameters that always have a definite numerical value in quantum mechanics—such as mass and charge—cannot be taken to refer to intrinsic properties of the particles. In general, anything that one may be inclined to regard as physical properties of the particles over and above their spatial configuration is not a property that can be considered as belonging to the particles taken individually, but is situated on the level of their quantum state. The quantum state, in turn, is defined on configuration space—that is, the mathematical space each of whose points represents a possible configuration of matter in physical space. Hence, again, the quantum state *presupposes* a configuration of objects given in terms of their location in physical space to which it is applied.

Nonetheless, one may envisage upholding the commitment to intrinsic properties by maintaining that parameters such as mass and charge are introduced in physics through the dynamical role that they play for the motion of the particles, but that the description in terms of a dynamical role refers to an underlying intrinsic property. Thus, Jackson (1998) writes,

> When physicists tell us about the properties they take to be fundamental, they tell us what these properties *do*. It does not follow from this that the fundamental properties of current physics, or of completed physics, are causal cum relational ones. It may be that our terms for the fundamental properties pick out the properties they do via the causal relations the properties enter into, but that at least some of the properties so picked out are intrinsic. They have, as we might put it, relational names but intrinsic essences. One way to block this result is holding that the nature of everything is relational cum causal, which makes a mystery of what it is that stands in the causal relations.
>
> (Jackson (1998), pp. 23–24)

The argument hence is that at least some of the dynamical parameters that a physical theory introduces have to be conceived as referring to intrinsic properties of the objects to which these parameters are attributed, because, otherwise, it would be a mystery as to what the objects are, whose dynamics is described by means of parameters such as mass or charge. This is an *a priori* argument in the vein of Aristotelian metaphysics according to which objects or substances must have an intrinsic essence or form (eidos). If physics does not deliver intrinsic properties, its descriptions in terms of dynamical relations among objects should nevertheless be conceived as referring to such properties. However, the problem with this argument is that, as mentioned earlier, dynamical parameters such as mass and charge presuppose objects given in terms of their spatial location to which they are applied so that these parameters are unsuitable to take the position of something that refers to an intrinsic essence of these objects. Moreover, it is a contingent matter of fact that in classical physics, such parameters are attributed to the objects taken individually; in quantum physics, they are situated on the level of their configuration. It is, therefore, worthwhile to look for something else than intrinsic properties that can do the job of individuating the objects that stand in the dynamical relations without falling into any sort of mystery.

Doing so does not automatically imply that one has to abandon the view of an intrinsic nature of the physical objects. Apart from intrinsic properties, two other main philosophical possibilities to uphold this view are available. The one possibility is to say that each particle has a primitive thisness—that is, a primitive individual essence known as a haecceity. This is not a qualitative feature: it is not constituted by any properties. It is the fact of being a certain individual object, whereby the fact of being that object is not tied to any characteristic features of the object. This is, therefore, a very problematic metaphysical stance, which has no basis in physics. It implies that there are ontological differences (permutation of haecceities) that do not make any physical difference.

The other possibility is to maintain that the particles are bare substrata, having a primitive stuff-essence (this view is usually traced back to Locke (1690), book II, ch. XXIII, § 2). By contrast to haecceities, this is not an individual essence, but a kind essence. Nonetheless, it remains unclear what a primitive stuff-essence could be, since it could not be characterized by any properties. In brief, thus, neither the idea of a primitive thisness nor the one of bare substrata is a convincing way out to uphold the view of the basic physical objects being constituted by an intrinsic essence, if physics blocks the Aristotelian way of taking the basic objects to be endowed with intrinsic properties. To paraphrase Jackson, a primitive thisness or a bare substratum make a mystery of what the objects are that stand in the dynamical relations described by parameters such as mass, charge and wave function.

What we then arrive at is a commitment to naked particles so to speak—that is to say, all there is to the particles is their spatial location. In fact, Democritus and Newton conceive the atoms as being inserted into an absolute background space, which is three-dimensional and Euclidean. However, the commitment to an absolute space is at least as problematic as the commitment to intrinsic physical properties of the particles. There are three central objections against this commitment:

1. *Differences that do not make a physical difference*: As Leibniz points out in his famous objections to Newton's substantivalism, there are many different possibilities to place or to transform the whole configuration of matter in an absolute space that leave the spatial relations among the material objects unchanged so that there is no physical difference between them (see, notably, Leibniz's third letter, §§ 5–6, and fourth letter, § 15, in Gerhardt (1890), pp. 363–364, 373–374, English translation Ariew (2000)).

2. *Infinite expansion*: Assume, as is well supported by physics, that the configuration of matter consists in finitely many discrete objects, such as point particles. If that configuration is embedded in an absolute space, then that space will stretch out to infinity, unless an arbitrary boundary is imposed (at least in a classical setting, since in general relativity the global matter distribution might determine a compact geometry); in any case, it will stretch out far beyond the actual particle configuration. However, all the experimental evidence is one of relative particle positions and change of particle positions—that is, motion. Thus, space is needed in physics only to describe the configuration of matter and, notably, the change in that configuration. Consequently, subscribing to the existence of an absolute space in which that configuration is embedded amounts to inflating the ontology.

3. *Problem of what fills space*: If one endorses a dualism of there being both matter and space as distinct entities, one faces the question of what the matter is that fills space (see Blackburn (1990)). What makes up the difference between a point of space being occupied and its being empty? However, none of the three possible answers of intrinsic properties, a primitive thisness or a bare substratum is convincing. The failure to come up with a cogent answer to this question provokes the objection that matter is inscrutable; this objection can then be turned into an argument against the very existence of matter—that is, an argument for idealism (see notably Foster (1982), ch. 4; cf. also Robinson (1982), ch. 7).

The impasse of not being able to put forward a characterization of matter that stands up to scrutiny arises only if one presupposes a dualism of an absolute space and matter as that which fills space. Relationalism, by contrast, is the view that there are only spatial relations—that is, distance relations—among physical objects, but no space in which these relations are embedded. Relationalism thereby avoids the three mentioned problems: all transformations that leave the distance relations among the material objects unchanged are mathematical representations of one and the same physical configuration, so that there are no differences in the ontology that do not make a physical difference. Furthermore, there is no space beyond the actual configuration of matter. Most importantly for our purposes, if there are only spatial relations among material objects, but no points of space, then these very relations are available to answer the question of what characterizes matter. In the following, we prefer the term "distance relations" to the term "spatial relations" in order to avoid the connotation of an underlying space.

To recap, our claim is that atomism, if set out in terms of point particles being inserted into an absolute space, fails to achieve the aim of being a parsimonious ontology. The consequence of this failure is that atomism, thus conceived, is unable to come up with a cogent answer to the question of what the atoms are. To answer this question, one has to abandon points of space and retain only point particles (matter points), with these point particles standing in distance relations that individuate them. The point particles then are simple in the sense that they have no parts or any other internal structure. In a nutshell, to accomplish the task of setting out atomism as a minimalist ontology of the natural world, only the following two axioms are required:

Axiom 1 *There are distance relations that individuate simple objects—namely, matter points.*
Axiom 2 *The matter points are permanent, with the distances between them changing.*

We submit that these two axioms are minimally sufficient to formulate an ontology of the physical world that is empirically adequate, given that all the empirical evidence comes down to relative particle positions and change of these positions. Hence, relationalism about space and time is motivated for us by the search for an ontology of matter that is most parsimonious while being empirically adequate.

Why do we single out the distance relations? Without acknowledging a plurality of objects in some sense—and be it so-called thin objects without intrinsic properties (see French (2010))—it is difficult to see, to say the least, how empirical adequacy could be achieved. If there is a plurality of objects, there has to be a certain type of relations in virtue of which these objects make up a configuration that then is the world. Generally speaking, one can conceive different types of relations making up different sorts of worlds. For instance, one may imagine thinking relations that individuate mental substances making up a world of minds. Lewis's hypothetical basic relations of like-chargedness and opposite-chargedness, by contrast, would not pass the test, since, as Lewis notes himself, these relations fail to individuate the objects that stand in them as soon as there are at least three objects (Lewis (1986a), p. 77).

When it comes to the natural world, the issue is relations that qualify as providing for extension. That is the reason to single out distance relations. In a future theory of quantum gravity, these relations may be conceived in a different manner than in our current and past physical theories. Nonetheless, we submit that relations providing for extension—namely, distances—are, given the state of the art in both physics and philosophy, the first choice for an ontology of the natural world that is to be empirically adequate. Change in these relations then is sufficient to obtain empirical adequacy. That is the reason to pose the two aforementioned axioms, and only these two ones. Of course, all these claims are fallible. To repeat, we are after minimal sufficiency; we make no claim about these axioms being necessary or *a priori*.

To convey what axiom 1 means, we have to choose a representation. Let us consider a universe consisting of a finite number of $N \in \mathbb{N}$ matter points. Taking the number of matter points to be finite is sufficient for empirical adequacy and will make the following discussion much easier. In order to obtain an ontology that is minimally sufficient for empirical adequacy, we assume that the set Ω of all possible configurations of distance relations between $N \in \mathbb{N}$ matter points can be represented as follows:

Definition 1 *Let* $\mathcal{M} = \{1, 2, ..., N\}$ *and* $\mathcal{E} = \{(i,j) \mid i,j \in \mathcal{M}, i \neq j\}$. *The set* Ω *comprises elements* $\Delta = (\Delta_{ij})_{(i,j) \in \mathcal{E}}$ *that can be represented by numerical assignments fulfilling the following requirements:*

  i. $\Delta = (\Delta_{ij})_{(i,j) \in \mathcal{E}}$ *is a* $\frac{N}{2}(N-1)$*-tuple of positive values* $\Delta_{ij} \in \mathbb{R}^+$ *for each* $(i,j) \in \mathcal{E}$.
  ii. *For all* $(i,j) \in \mathcal{E}$ *one has* $\Delta_{ij} = \Delta_{ji}$.
  iii. *For all* $i,j,k \in \mathcal{M}$, *it is the case that* $\Delta_{ij} \leq \Delta_{ik} + \Delta_{kj}$.
  iv. *For all* $i,j \in \mathcal{M}$, *if* $i \neq j$, *then* $\{\Delta_{ik} \mid k \in \mathcal{M}, k \neq i\} \neq \{\Delta_{jk} \mid k \in \mathcal{M}, k \neq j\}$.

Requirements (i) and (ii) state that the distance relation is, in mathematical terms, irreflexive, symmetric and connex (meaning that all matter points in a given configuration must be related with one another). Requirement (iii) is the triangle inequality in virtue of which the relation in question is a distance relation. Requirement (iv) states that the distance relations individuate the matter points: if matter point i is not identical with j, then the two sets that list all the distance relations in which these points stand with respect to all the other points in a configuration must differ in at least one such relation. It is such differences in the way in which i and j relate with the other points in the configuration that make it that i and j are different matter points. Consequently, by means of (iv), we avoid having to accept the plurality of matter points as a primitive fact, which would imply that the matter points are bare particulars or bare substrata. Instead, they are individuated by the distance relations.

Formulating these requirements in terms of numerical assignments is a means to express the features in virtue of which a relation is a distance relation. A minimal requirement in that respect is the triangle inequality (iii).

Furthermore, the ratios between the distances are part of the ontology (so that the distances can individuate the matter points). By contrast, the numerical assignments do not belong to the ontology, let alone the notion of absolute scale that comes with them; they are just introduced for representational purposes. Consequently, the fact that the values assignable to Δ_{ij} are real numbers does not smuggle in any infinity in this ontology: there is a finite number of N matter points and, hence, finitely many distance relations. Contrary to a well known and long debated argument by Field (1980), in using real numbers, we are not committed to mathematical Platonism in having to subscribe to an ontological commitment to real numbers. In general, in using something as a representational means—numbers, space and time, dynamical parameters, laws, etc.—one is not committed to endorsing that means in one's ontology. Doing so requires an argument. The minimalist in ontology sets out to rebut any such argument by claiming that (a) a minimalist ontology as given by the two aforementioned axioms is sufficient as truth-maker for all the true propositions about the world and that (b) additional ontological commitments do not amount to a gain in explanation, but to commitments to surplus structures that entail drawbacks.

Nonetheless, by introducing a labelling $\mathcal{M}$ of the matter points, this definition can be taken to suggest that their numerical plurality is a primitive fact. But this is just an artifact of the set-theoretical language. The elements of a set $\mathcal{M}$ *qua* set-theoretical objects have to be numerically distinct for $\mathcal{M}$ to be a well-defined set of N objects. However, the referents of this formalism—the matter points *qua* physical objects—are individuated by the distance relations given by Δ, so that these relations account for their numerical plurality.

To emphasize the permutation invariance of the matter points in the formalism, it is possible to make the earlier definition of Ω independent of the labelling by introducing the following equivalence relation:

Definition 2 *Take $\Delta, \Delta' \in \Omega$, and consider $\mathbb{S}_N$ as the set of all possible permutations of elements of $\mathcal{M}$. We define $\Delta \simeq \Delta'$ if and only if there is a permutation $\sigma \in \mathbb{S}_N$ and a constant $c \in \mathbb{R}^+$ such that for all $(i, j) \in \mathcal{E}$ it is the case that $\Delta'_{ij} = c\Delta_{\sigma(i)\sigma(j)}$.*

The set

$$\tilde{\Omega} = \Omega/\simeq := \{[\Delta]_\simeq \,|\, \Delta \in \Omega\}, \qquad [\Delta]_\simeq = \{\Delta' \in \Omega \,|\, \Delta' \simeq \Delta\} \qquad (2.1)$$

then comprises all possible configurations of distance relations independently of a labelling of the matter points and a scaling. The latter implies that the relevant quantities are not the absolute values of the distance relations Δ_{ij}, but only the ratios Δ_{ij}/Δ_{kl}.

One way to envision an element of $[\Delta]_\simeq \in \tilde{\Omega}$ is by a representative $\Delta \in \Omega$ that can be viewed as a coloured graph $G(\Delta) = (\mathcal{M}, \mathcal{E}, \Delta)$ in which $\mathcal{M}$ are the nodes, $\mathcal{E}$ are the edges, and to each edge $(i, j) \in \mathcal{E}$ the colour Δ_{ij} is attached. Also the graphs $G(\Delta)$ can be made label-independent by considering the equivalence classes $[G(\Delta)]_\simeq = \{G(\Delta')|\Delta' \simeq \Delta\}$ and treating $G(\Delta)$ only as the corresponding representative of the class.

Accordingly, distances individuating point-objects that then are matter points and change of these distances are the primitives of the minimalist ontology that we propose. Indeed, using the material terms of unextended substances being at a distance from each other with that distance changing is ontologically more accurate to characterize this position than the formal categories of objects and relations (cf. Mulligan (2012)). That notwithstanding, we will continue to talk in terms of objects and relations to link up with the literature.

In posing distance relations instead of an absolute space, we follow Leibniz's relationalism about space. According to Leibniz, distances make up the order of what coexists (third letter, § 4, in Gerhardt (1890), p. 363). Distances are able to distinguish objects, thus respecting Leibniz's principle of the identity of indiscernibles (although Leibniz himself does not have a relational individuation of objects in mind). They thereby account for there being a plurality of objects. Consequently, we do not presuppose the numerical plurality of objects as given, which would be objectionable: there would then be nothing which makes it that there is more than one object. Quite to the contrary, in virtue of the matter points standing in distance relations that distinguish them, there is a plurality of them. In other words, in virtue of these relations, there is a configuration of matter points that is constituted through *variation* in the distance relations that connect the matter points and that make it that these are matter points. This conception furthermore accounts for the impenetrability of matter without having to invoke a notion of mass: for any two matter points to overlap, it would have to be the case that there is no distance between them.

Hence, matter is structurally individuated—namely, by the distances among the material objects. In other words, *we conceive relationalism as a form of ontic structural realism, employing the relations that relationalism about space acknowledges to individuate the relata of these relations.* As the literature on ontic structural realism has made clear, structures in the sense of concrete physical relations—such as distances—can individuate physical objects (see, e.g., Ladyman (2007)). Indeed, structures in this sense can do exactly the same as properties are supposed to do: if one holds that objects are bundles of properties, then the corresponding view is what is known as *radical* ontic structural realism—namely, the view that objects are constituted by relations, being the nodes in a network of relations (see Ladyman and Ross (2007), ch. 2 and 3, and French (2014), ch. 5–7, for an elaborate defense). If one thinks that there are underlying substances that instantiate properties, then the corresponding view is what is known as *moderate* ontic structural realism—namely, the view that objects and relations are on a par, being mutually ontologically dependent: relations require relata in which they stand, but all there is to the relata is given by the relations that obtain among them (see Esfeld (2004), section 3; Esfeld and Lam (2011); this view is taken up in, e.g., Floridi (2008) and McKenzie (2014)).

In any case, the fundamental objects do not have an intrinsic nature, but a relational one. Relations are on the same footing as intrinsic properties in that respect. To come back to the citation from Jackson (1998) earlier, if it were mysterious what it is that stands in the relations (on the assumption that all there is to a fundamental physical object are the relations in which it stands to other such objects), then it would be mysterious in exactly the same way what it is that instantiates the intrinsic properties that are supposed to characterize the fundamental physical objects. To put it differently, in any case, bare particulars are mysterious, and the commitment to bare particulars is avoided by taking certain intrinsic properties, or certain relations to be essential for the fundamental physical objects.

We advocate moderate ontic structural realism. To our mind, there is no physical or metaphysical reason to conceive ontic structural realism as being opposed to an object-oriented metaphysics: if there are relations, there are objects that stand in the relations. In other words, ontic structural realism can admit objects, as long as all there is to the objects are the relations among them. What ontic structural realism rejects is the property-oriented metaphysics that dominates philosophy since Aristotle: the fundamental physical objects do not have an intrinsic essence. This is a conception of objects that stands on its own feet, being an alternative to both the view of objects as bare substrata and the view of objects as bundles of properties (or relations). There is a mutual ontological dependence between objects and relations: as there cannot be relations without objects that stand in the relations, so there cannot be objects without relations in which they stand. Hence, if one removed the distance relations, there would not remain bare substrata, but there would then be nothing (see Esfeld and Lam (2008, 2011)).

In order to obtain the result that the distance relations individuate the matter points and thus distinguish them, we have to stipulate that these relations establish what is known as absolute discernibility in today's literature: each of the matter points distinguishes itself from all the other ones by at least one distance relation that it bears to another matter point (see requirement (iv) in definition 1). Being absolutely discernible, the matter points are individuals. Hence, the famous slogan "No entity without identity" coined by Quine (1969, p. 23) applies to the matter points, although there is nothing intrinsic about them: they do not have an intrinsic identity, but one provided by distance relations that make them absolutely discernible entities.

What is known as weak discernibility in today's literature would not be enough, since weak discernibility does not avoid having to endorse a given numerical plurality of objects: for weak discernibility to be satisfied, it is sufficient that objects stand in an irreflexive relation, without there being anything that distinguishes one object from the other ones. Hence, weak discernibility indicates that there is a numerical plurality of objects, but is too weak to individuate the objects.[1]

Nonetheless, there is a tension in ontic structural realism between the following two claims:

1.   The symmetries that physical theories exhibit are the guide to the ontic structures.
2.   The ontic structures individuate physical objects.

The tension consists in the fact that the higher a degree of symmetry a structure exhibits, the less it is in the position to individuate objects (see Keränen (2001)). In posing axiom (1), we propose to resolve this tension by simply dropping the first of these two claims. What we need structures in a parsimonious ontology for is to relate simple objects in such a way that the structures individuate the objects. In other words, the rationale for admitting structures in the ontology rests on the second of these two claims.

Indeed, any model satisfying axiom (1) has to include at least three matter points and has to comply with requirement (iv) of definition 1. Symmetrical configurations are thus ruled out, as well as, for instance, the configuration of an isosceles triangle. However, this is no objectionable restriction: having empirical adequacy in mind, there is no need to admit worlds with only one or two objects or *entirely* symmetrical worlds as physically possible worlds (and see Hacking (1975) and Belot (2001) for an argument not to admit these as metaphysically possible worlds either). The restriction formalized by (iv) does not rule out universal configurations that are *locally* symmetric (i.e. having only a subset of matter points in a symmetric configuration). What is excluded are cases in which *all* the matter points in a universal configuration are arranged symmetrically. Having in mind empirical adequacy, we note that this restriction poses no problem for the ontology of physics: our universe definitely is not in a symmetric configuration. That notwithstanding, symmetries in physical theories are very important to achieve a description of the evolution of our universe that is both simple and informative, as we shall explain later: they entail a great improvement in simplicity with only a little loss in information about the actual particle configuration of the universe (which is not symmetric). In short, the benefit of requirement (iv) is that it entitles us to conceive the distance relations as individuating the matter points, thus avoiding the commitment to bare particulars; the cost is, in fact, none since (iv) by no means calls the significance of symmetries in physical theories for the description of the evolution of the universe into question.

Furthermore, there is no need to abandon absolute discernibility in quantum physics either (since Bohmian mechanics solves the measurement problem by, among other things, respecting the individuality of the quantum particles; we will go into Bohmian mechanics in Chapter 3, section 2). In brief, its position in the network of distance relations distinguishes each matter point from all the other ones so that the matter points are absolutely discernible, but they are permutation invariant in the sense that labelling the matter points has no significance. To sum up this important issue: to obtain objects that are structurally individuated by the relations in which they stand, we need absolute discernibility; to obtain absolute discernibility, we have to throw symmetries in the sense of entirely symmetrical, global configurations out of the ontology; doing so makes the ontology empirically adequate (since the actual configuration of matter of the universe is not symmetrical) without infringing upon the representational importance of symmetries.

Since all there is to the matter points are the distance relations in which they stand, this is, like Cartesianism, a geometrical conception of matter. However, it is not to be confused with super-substantivalism—that is, the view that space (or space-time) is the only substance and matter a property of space (see most recently Lehmkuhl (2016)). The main problem for this view is to account for motion, if there are only points of space (or space-time) and their topological and metrical properties, since these cannot move. Indeed, Wheeler (1962) tried super-substantivalism out in his pro-gramme of geometrodynamics, but failed in the attempt to reduce dynami-cal parameters to geometrical properties of points of space-time (see Misner et al. (1973), § 44.3–4, in particular, p. 1205). By contrast, if there are no points of space or space-time, but only distance relations between sparse points that hence are matter points, all the dynamical parameters that figure in physical theories can then be construed in terms of the role that they play in accounting for the change in these relations—that is, the motion of these points. In short, in a geometrical conception of matter by distance relations between sparse points, there is a clear sense in which there is motion and dynamical parameters capturing motion.

However, the substantivalist who accepts a dualism of matter and space can retort that by endorsing an absolute space that underlies the spatial con-figuration of matter, the substantivalist ontology, although being less simple than the relationalist one, gains in explanatory value. Thus, Maudlin (2007a, pp. 87–89) takes length of a path in absolute space as the primitive notion and derives the notion of distance of point particles from that notion as the minimal path length connecting them, claiming that he is thereby able to explain the constraints on the distance relation (such as the triangle inequality). The concern, however, is that one does thereby not provide a deeper explanation of the distance relation: in order to be able to define a minimal path length in space, one has to presuppose a structure that is rich enough to accommodate a metric—as the relationalist has to presup-pose a relation that is rich enough to fulfill the triangle inequality in order to count as distance relation. If one employed a primitive notion of path that does not permit a definition of minimal length, then one could not derive the distance relation from such a notion of path. In short, sub-stantival space comes with a metric in terms of, for example, paths of geodesic motion, and any metric defining a physical space is such that it fulfills all the constraints of three-dimensional geometry. Hence, there is no additional explanatory value here in comparison to the relationalist who just presupposes that the relations admitted as primitive fulfill certain requirements; there only is the disadvantage that substantival space contains more structure than is needed to account for the empirical evidence, which consists of relative particle positions and change of these positions. In a nutshell, the substantivalist creates the illusion of giving a deeper explanation of something that, actually, comes in a package with the postulation of a substantival space.

Since the distance relations individuate the matter points, this view spells out what Schaffer (2010a) calls "the internal relatedness of all things", because certain relations—namely distances—are constitutive of the fundamental physical objects. However, this view does not end up in monism (*pace* Schaffer (2010b)): there is a plurality of objects—namely, a plurality of matter points being at a distance of each other. Indeed, the matter points can with good reason be considered as substances: although they are not independent, since being tied together by distance relations is essential for them, they are permanent. They do not come into existence, and they do not go out of existence. The attractiveness of atomism depends on being able to vindicate the matter points as substances in this sense: if their number did not stay constant, atomism would be committed to the absurd view that matter points come into existence out of nothing and are literally annihilated, disappearing into nothing. As Parmenides famously maintains,

> What-is is ungenerated and imperishable, a whole of one kind, unperturbed and complete. Never was it, nor shall it be, since it now *is*, all together, one, continuous. For what birth would you seek for it? Where, whence did it grow? Not from what-is-not will I allow you to say or to think; for it is not sayable or thinkable that it is not. And what need would have stirred it later or earlier, starting from nothing, to grow? Thus it must be completely or not at all.
>
> (fragment Diels-Kranz 28 B8, quoted from Graham (2010), pp. 215–217)

On atomism, what-is are the matter points (particles). They are therefore substances.

As the matter points are permanent, so is the change in their distance relations. To invoke again Parmenides, we noted that being is ungenerated and imperishable, and change belongs to being. That is to say, as the matter points do not come out of nothing and do not disappear into nothing, so change does not arise out of a sudden and does not suddenly cease to occur. If there are only matter points connected by distances, all change is change in the distance relations among the permanent matter points.

Let us therefore now turn to *axiom 2*. Change, conceived as change in the distances among matter points that are permanent, does not presuppose any temporal notion. Nonetheless, change, thus conceived, is directed in the following sense: it goes from one particular state of the configuration of matter points consisting in certain distances among the matter points to another particular state of that configuration consisting in other distances among some matter points. Any such change may be reversible. Nevertheless, the actual change in the configuration is directed by virtue of the fact that it goes from one specific state of the configuration to another specific state of the configuration. By contrast, there is no direction in the distance relations individuating matter points as given by axiom 1, since there is no spatial direction as long as there are only distance relations, but no space into which these relations are embedded. Thus, in the ontology that we propose, matter points, distance relations and the change of these relations are the only primitives.

If change is change in the distance relations, any change in any possible model of this ontology has to be such that it satisfies the requirements stated in the definition of the distance relation—that is, definition 1. In particular, by axiom 1 and requirement (iv), the distance relations individuate the matter points and thereby establish their identity in any state of the configuration of matter points. Given that, as on axiom 2, the matter points are permanent, the change in the distance relations provides for the identity of the matter points in the configuration across all change. In other words, what makes (i) the same matter point through each state of the configuration is the way in which the relations it bears with respect to all the other matter points change. An appropriate manner to represent this identity, therefore, consists in depicting the matter points as moving on continuous trajectories.

One may object that, if we are confronted with two separate snapshots depicting two distinct states of the universal configuration of matter points, then it might be impossible for us to identify the matter points across the snapshots (that is, there is, in general, no unique way to order the snapshots and connect them by means of continuous trajectories), *contra* our individuation claim. However, the source of this impasse is epistemological rather than ontological, since it just stems from the fact that no information is provided about the change that led from one snapshot to the other. Once we have this information, no individuation worry arises, because the distance relations and their change provide for the identity of the matter points.

Against this background, the change in the distance relations (and, hence, the motion of the matter points) can be represented as a parametrized list of states of the configuration of matter—that is, a map

$$\Delta_{(\cdot)} : \mathbb{R} \to \tilde{\Omega}, \qquad \lambda \mapsto [\Delta_\lambda]_{\simeq}, \tag{2.2}$$

which, by means of $\lambda \mapsto [\Delta_\lambda]_{\simeq}$, denotes the change of the distance relations in the configuration of matter independently of a labelling. Again, by axiom 1, it is ruled out that the universal configuration of matter points can evolve into a state that violates requirement (iv), such as the evolution into a symmetrical *universal* configuration (whereas the evolution into *local* symmetrical configurations is by no means excluded). However, again, this does not deny the importance of symmetries in physical theories in order to accomplish a simple representation of the change that actually occurs: although the universe is by no means in a symmetrical state, employing symmetries leads to a huge simplification in the description of the evolution of the universe (and its subsystems). Banning symmetrical universal configurations from the ontology does not infringe upon the empirical adequacy of the ontology and allows us to employ the distance relations and their change to individuate the matter points in any given universal configuration and to provide for their identity across change.

As the minimalist ontology that we propose does not imply absolutism about space, so it does not imply absolutism about time: time derives from change. Again, we follow Leibniz for whom time is the order of succession (see notably Leibniz' third letter, § 4, and fourth letter, § 41 in Gerhardt (1890), pp. 363, 376). Hence, there is no time without change, but change exhibits an order, and what makes this order temporal is that it is unique and has a direction. This in particular implies that the notion usually referred to as the "identity over time" of a particle is supervenient on the distance relations and their change.

Although Leibnizian relationalism thus entails that the topology of time induced by the unique ordering of the elements in $\tilde{\Omega}$ is absolute, this is perfectly consistent with relationalism about time, since time depends on the change in the configuration of matter points. In Mach's words, "time is an abstraction, at which we arrive by means of the change of things" (Mach (1919), p. 224). The important point is that there is no external measure of time: the idea that the global dynamics unfolds according to the ticking of a universal clock is meaningless. If the entire universe could evolve at different external time rates, then two such evolutions would be physically indistinguishable, given that they would consist in the very same sequence of states of the universal configuration of spatial relations. Hence, they would exhibit the very same change in the distances among the matter points. In short, a commitment to an universal external clock would introduce ontological differences that would not make any physical difference (this point is made also in, e.g., Barbour and Bertotti (1982), see especially pp. 296–297). Consequently, there is no absolute metric of time. What we call "time" in this context is just an arbitrary parametrization of the curve $\lambda \mapsto [\Delta_\lambda]_\simeq$ on $\tilde{\Omega}$ and not, as in the Newtonian case, an additional external variable. For this reason, the only meaningful way to define a clock is to choose a reference subsystem within the universe relative to which the rate of change in distance relations is measured. An example of a simple reference subsystem is the circular motion of a pointer on a dial of a watch, the arc length drawn by the pointer being directly related to the parameter λ in the definition of the map $\lambda \mapsto [\Delta_\lambda]_\sim$.

To repeat, the argument for the ontology being exhausted by the two axioms of there being distance relations individuating matter points and these distances changing is that one thus achieves a minimalist ontology of the physical world that is empirically adequate and that any richer ontology does not lead to a gain in explanation, but only to new drawbacks. However, we obtained this ontology by starting from the paradigm of atomism, working through an argument to the effect that atomism, when elaborated on as a cogent ontology of matter, comes down to only these two axioms. But atomism is not the only proposal for a fundamental ontology of the physical world. The opposite view is to take matter to be one continuous stuff, known as gunk, that fills all of space (note that this view does not have to commit itself to points, neither matter points nor points of space; see Arntzenius and Hawthorne (2005)). This view also goes back to the first Presocratic natural philosophers: Thales, Anaximander and Anaximenes can all be taken to embrace a stuff view of matter. Let us call this view "ontological monism", because it admits only one continuous substance, to mark the contrast with the ontology of atomism, according to which there is a plurality of discrete substances. Like atomism as set out in this section, this view is not tied to endorsing an absolute background space: it can be construed as being committed to a continuous stuff that is extended, but not to an absolute space that is distinct from that stuff and into which that stuff is inserted.

In order to accommodate variation, gunk cannot be conceived as being homogeneous throughout space. To take variation into account, one has to maintain that there is more stuff in some parts of space and less stuff in others. Atomism conceptualizes variation in terms of different distances among the discrete matter points so that some matter points are situated close to one another, whereas others are further apart: there are clusters of matter points with distances among them that are smaller than the distances that these matter points bear to matter points outside such a cluster. By contrast, the gunk ontology cannot accommodate variation throughout space in terms of the concentration of primitive matter points. It therefore faces the following question. What constitutes the fact of there being more matter in some regions of space and less matter in others? Consider what Allori et al. (2014) say in this respect when discussing the proposal for a primitive ontology of quantum physics in terms of a continuous matter density that fills all of space, known as the GRWm theory (we will go into this theory in Chapter 3, section 3):

> Moreover, the matter that we postulate in GRWm and whose density is given by the *m* function does not ipso facto have any such properties as mass or charge; it can only assume various levels of density.
>
> (Allori et al. (2014), pp. 331–332)

Hence, the view of matter being gunk has to acknowledge as a further primitive a variation of the density of gunk throughout space with gunk being more dense in some parts of space and less dense in other parts. That is to say, the primitive stuff admits of degrees, as expressed by the m function in the GRWm formalism: there is more stuff in some parts of space than in others, with the density of matter in the parts of space changing in time; otherwise, the theory would not be able to accommodate variation. Formally, one can represent the degrees of density in terms of attributing a value of matter density to the points of space (the m function as evaluated at the points of space), although the matter density stuff, being gunk, is infinitely divisible, and this ontology is not committed to the existence of points of space. The main problem is that it remains unclear what could constitute the difference in degrees of stuff at points of space, if matter just is primitive stuff. The gunk theory thus is committed to the view of matter being a bare substratum with its being a primitive fact that this substratum has various degrees of density in different parts of space. In a nutshell, there is a primitive stuff-essence of matter that furthermore admits of different degrees of density.

In comparison to the gunk view of matter, atomism is the simpler and clearer proposal for a fundamental ontology, because it avoids the dubious commitment to a bare substratum or primitive stuff-essence of matter with different degrees of density. Apart from this *a priori* argument, its empirical success speaks in favour of atomism. All experimental evidence in fundamental physics is evidence of relative particle positions and change of these positions. Entities that are not particles—such as waves or fields—come in as figuring in the explanation of the motion of the particles, but they are not themselves part of the experimental evidence. Moreover, science in general teaches us that macroscopic objects are composed of chains of molecules and that these molecules are composed of atoms in the sense of the chemical elements of the periodic table, which in turn are composed of protons, neutrons and electrons. The ontology of atomism can easily account for that composition: we finally get down to matter points standing in distance relations. Everything else is composed of these simple, discrete objects, consisting in certain clusters of matter points. That is how we get from fundamental particle physics to statistical physics of large ensembles of elementary particles, chemistry, molecular biology and, finally, neuroscience. And that is why Feynman says at the beginning of the *Feynman lectures on physics*, as quoted in the introduction, that the atomic hypothesis contains an enormous amount of information about the world (Feynman et al. (1963), ch. 1–2).

Nonetheless, atoms *qua* matter points are theoretical entities. They are not seen by the naked eye when one sees, for instance, dots on a screen as outcomes of the double slit experiment in quantum mechanics. They are admitted because they provide the best explanation of the observable facts. The simplicity and parsimony of this proposal are part of the case for its being the best explanation. To put it in a nutshell, *particle evidence is best explained in terms of particle ontology*. However, this explanation is not given by the ontology of matter points alone, but by this ontology together with the dynamics that is put forward to describe the motion of the matter points: it is the dynamics that provides for the stability of the macroscopic objects with which we are familiar. For instance, an ephemeral cat-shaped configuration of matter points would not be a cat; only a stable such configuration is a cat. To obtain this stability, as we will show in the following, no properties of the matter points over and above the distances among them are needed, but a dynamics is that provides for trajectories such that there are stable macroscopic configurations of atoms.

To conclude this section, let us come back to the term "primitive ontology" that is used in the context of quantum physics to refer to the distribution of matter in physical space whose state the quantum formalism describes (cf. Dürr et al. (2013b), ch. 2). We propose to endow this term with a threefold substantial meaning:

1. The ontology is primitive in the sense of *fundamental*: the matter points are not composed of anything, but they compose everything else.
2. It is primitive in the sense of referring to *primitive objects*: the matter points have no intrinsic properties. However, they are not bare substrata either. The distance relations in which they stand are their essence.
3. It is primitive in the sense of *factual*: the configuration of matter points is simply there. The matter points have no further function, whereas everything else—the geometry of space-time, dynamical parameters such as mass, charge, energy, etc.—is introduced through the function that it has for the evolution of the configuration of matter points (that is, its function for the change in the distances among the matter points).

Consequently, we go beyond the term "primitive ontology" as it is usually employed in the literature—namely, in the sense of the primitive ontology of a given theory (cf. Allori et al. (2008))—by proposing a primitive ontology *tout court*. However, this fundamental ontology is not the foundation of knowledge. As mentioned earlier, matter points are theoretical entities. The justification of this ontology is a coherentist one: matter points as fundamental objects and a dynamics for the evolution of the distances individuating these objects is the way to achieve an overall coherent system of our knowledge that provides for a clear and elegant explanation of the experimental evidence (in terms of laws, causes and/or unification).

2.2   The account of change: why simplicity in ontology and simplicity in representation pull in opposite directions

So far we have been concerned with simplicity in the sense of parsimony in ontology: distance relations individuating matter points, with these relations changing, while the matter points are permanent, is the first and foremost candidate for the simplest ontology of the natural world that is coherent and empirically adequate, given that all the experimental evidence comes down to relative particle positions and their change. However, when it comes to representing that change in a physical theory, the conceptual means provided by this ontology are insufficient. Using only these conceptual means, we could not do much better than just listing the change that actually occurs, but not formulate a simple law that captures that change.

The rationale for seeking for a law is simplicity in representation: in the ideal case, the law is such that given an initial configuration of matter as input, the law yields a description of all the—past and future—change of the configuration as output. In this case, we have optimized both the simplicity and the informational content of the representation: specifying an arbitrary initial configuration of matter and putting the representation of that configuration into the law, we obtain the complete information about its change. However, in order to achieve such a law, we need more parameters than distances individuating matter points. The reason is that there is nothing about the distance relations in any given configuration of matter that provides information about the—past and future—evolution of these relations. That is why further parameters—both geometrical and dynamical ones, over and above relative distances that change—have to be attributed to the configuration of matter points to obtain a dynamical law.

Consequently, simplicity in ontology and simplicity in representation pull in opposite directions. Using only the concepts that describe what there is on the simplest ontology (matter points individuated by distance relations), the description of the evolution of the configuration of matter would not be simple at all, since one could not do much better than merely listing that change instead of capturing it in a simple and general law. Reading one's ontological commitments off from the simplest description—such as, e.g., Newtonian mechanics—the ontology would not be simple at all: it would in this case be committed to absolute space and time, to momenta, gravitational masses, forces, etc. A similar observation applies to the case of symmetrical universal configurations of matter mentioned in the preceding section: working with such configurations increases simplicity in representation. However, if one admitted such configurations in the ontology, one would lose the elegance and parsimony of the ontology, since, in this case, the matter points could not be individuated by the relations in which they stand, but would have to be conceived as bare particulars.

In general, our claim is that whatever geometry and dynamics a physical theory employs, all this apparatus is there only as the means to achieve a description of the change in the distance relations that is optimal with respect to both simplicity and informational content about that change. Hence, all the mathematical framework that we introduce here is only a means, limited to our human abilities, to efficiently describe that change. In this spirit, the apparatus we lay out in the following is surely not the only possibility to represent change. Indeed, there may even be better suited methods to do so. Furthermore, one description may only hold in a certain physical regime and has to be adapted or developed anew to include previously unknown physical phenomena, while the ontology remains the same.

In this section, we set out the apparatus that we employ to represent the change in the distance relations in general terms before discussing concrete physical theories from Chapter 3 on. In the next section, we elaborate on the philosophical position that allows us to both maintain minimalism in ontology and to endorse the simplicity in representation that comes with the geometry and dynamics that a physical theory uses.

Hitherto, we have formalized the mathematical counterparts of the elements of the ontology: the matter points, their distances, and the change thereof. To achieve the formulation of a physical law that describes that change, we need an additional structure that allows us to represent how the change, encoded in the curve $\lambda \mapsto [\Delta_\lambda]_\sim$, occurs. One mathematical framework to formulate such a law that has proven itself extremely powerful is calculus. A physical law can then be given by means of a velocity field from which the entire change—that is, $[\Delta_\lambda]_\sim$ for all $\lambda \in \mathbb{R}$—can be inferred uniquely by specifying an initial configuration $[\Delta_0]_\sim$ only.

One way to achieve a description through calculus is to find a differentiable manifold S equipped with a metric d and one strictly monotonic map $\tau : \mathbb{R} \to \mathbb{R}$ such that, for each $\lambda \in \mathbb{R}$, the configuration $[\Delta_\lambda]_\sim$ with $\Delta_\lambda = (\Delta_{ij,\lambda})_{(i,j) \in \mathcal{E}}$ can be represented in S by means of N points $q_{1,t}, q_{2,t}, ..., q_{N,t} \in S$ parametrized by $t \in \mathbb{R}$ that fulfill, for all $(i,j), (k,l) \in \mathcal{E}$ and $t \in \mathbb{R}$,

$$\frac{d(q_{i,t}, q_{j,t})}{d(q_{k,t}, q_{l,t})} = \frac{\Delta_{ij,\tau(t)}}{\Delta_{kl,\tau(t)}}. \tag{2.3}$$

This requirement stipulates that it must be possible to specify an embedding of $\lambda \mapsto [\Delta_\lambda]_\sim$ in at least one geometry in form of a differentiable manifold such that (i) the value $d(q_{i,t}, q_{j,t})$ at each edge $(i,j) \in \mathcal{E}$ represents the distance relation $\Delta_{ij,\tau(t)}$ between two matter points $q_{i,t}$ and $q_{j,t}$ in S, and (ii) the change in the distance relation $\Delta_{ij,\tau(t)}$ is represented by the change of the distance $d(q_{i,t}, q_{j,t})$. This representation hence requires the introduction of many new degrees of freedom that are not part of the ontology—namely (1) the scaling function of time τ, (2) the particular labels given to the matter points $q_{1,t}, q_{2,t}, \cdots, q_{N,t}$ and (3) the structure that comes along with the introduction of S and d.

Because of its simplicity as a representational means and its empirical adequacy in the non-relativistic regime, we shall frequently employ the choice of three-dimensional space $S = \mathbb{R}^3$ with d being the Euclidean distance as an example. Thus, by embedding the configuration of distance relations of matter points into the geometry of Euclidean space $S = \mathbb{R}^3$, we introduce a particular choice of orientation, rotation, translation, and scaling of the coordinate axes and by means of the vector space structure also directions $q_{i,t} - q_{j,t}$. However, to stress again, this is only a means of representation: the configuration of distance relations $[\Delta]_\sim$ is by definitions 1 and 2 in the preceding section independent of these additional degrees of freedom.

If such a geometrical embedding is possible, the purely relational change $\lambda \mapsto [\Delta_\lambda]_\sim$ can be cast as change with respect to the embedding manifold S in terms of the map

$$Q_{(\cdot)} : \mathbb{R} \to S^N, \qquad t \mapsto Q_t = (q_{1,t}, \cdots, q_{N,t}). \qquad (2.4)$$

The additional degrees of freedom introduced when choosing a representation make it possible to encode change by means of a derivative as a velocity field:

$$v_t(Q_t) := \frac{d}{dt} Q_t, \quad \text{for} \quad t \mapsto Q_t, \qquad (2.5)$$

provided the motion (2.2) and the chosen representations are sufficiently smooth to allow for the evaluation of the derivative. The velocity field v_t assigns to each point $Q_t \in S^N$ an element in the tangent space of $T_{Q_t} S^N$— that is, the space of directions in which change may occur. Given a certain motion $t \mapsto Q_t$ we could then infer the velocity field v_t. However, if we knew v_t as a map $S^N \to TS^N$ beforehand, all possible $t \mapsto Q_t$ could be inferred uniquely from it as integral curves of v_t starting at a certain initial value Q_0.

It is then the task of physics to find a general law of motion in terms of v_t that represents the change in the distances among the matter points in an efficient manner. Often this is done by induction on the basis of observed velocity fields (2.5) that exemplify the general law well in certain regimes. Obtaining a possible v_t by means of a physical theory thus achieved, the velocity field should also apply to more general motions. Hence, one can turn the definition in (2.5) around and study all motions $t \mapsto Q_t$ that fulfill

$$\frac{d}{dt} Q_t = v_t(Q_t). \qquad (2.6)$$

As already indicated, many degrees of freedom enter into such a representation. To make this explicit in a simple example, consider a possible representation of $\lambda \mapsto [\Delta_\lambda]_\sim$ in Euclidean space $\mathcal{S} = \mathbb{R}^3$. For $t = \tau(\lambda)$ and $\lambda \in \mathbb{R}$ such a representation can be achieved as follows. In a first step, one arbitrarily chooses one matter point, let us call it the N-th matter point, and represents it by $q_N = 0$. In a next step one has to choose an orientation of the Euclidean coordinate axes and may then represent the rest of the matter points inductively as $q_{1,t}, \dots, q_{N-1,t} \in \mathbb{R}^3$ such that the Euclidean distances $d(q_{i,t}, q_{j,t}) = |q_{i,t} - q_{j,t}|$ fulfill (2.3), which allows for an arbitrary scaling. The involved choices in the representation are (a) the labeling of the matter points and the origin of the coordinate system and (b) the orientation of the coordinate axes. Shifting the origin from one matter point to another and choosing a different orientation of the axes in one time instant amounts to the—possibly time dependent—translations and rotations, respectively.

In general, for a given representation $\mathcal{S}$, one may retrieve from a representation $t \mapsto Q_t$ the actual configuration of distance relations $\lambda \mapsto [\Delta_\lambda]_\sim$ by modding out these introduced degrees of freedom of the representation. This shall be denoted by the map

$$\Pi : \mathcal{S}^N \to \tilde{\Omega} \tag{2.7}$$

such that, for all $t \in \mathbb{R}$, it holds that $\Delta_{\tau(t)} = \Pi(Q_t)$. The change $\lambda \mapsto [\Delta_\lambda]_\sim$ is then uniquely identified by $t \mapsto Q_t$. Different choices in the degrees of freedom, however, usually imply a change in the velocity field v_t as it depends on the differential structure of $\mathcal{S}^N$ and, hence, the choices that had to be made to introduce this structure.

As an example, let us consider the case of two different representations of $\lambda \mapsto [\Delta_\lambda]_\sim$ in terms of $t \mapsto Q_t$ and $t \mapsto Q'_t$ in one and the same $\mathcal{S}$. Let us further denote their relationship by a time-dependent one-to-one map $T_t : \mathcal{S}^N \to \mathcal{S}^N$ such that $Q'_t = T_t(Q_t)$. Then the relationship between the velocity fields v_t guiding Q_t and v'_t guiding Q'_t, respectively, must fulfill the following diagram (Figure 2.1):

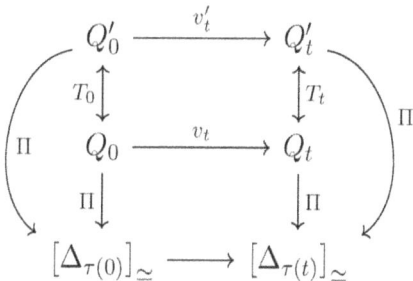

Figure 2.1 Different representations of the same motion.

This diagram highlights that the fundamental state of affairs is encoded only in the change of relational configurations $\lambda \mapsto [\Delta_\lambda]_\sim$; the way we "lift up" such a change to $\mathcal{S}^N$ is somewhat arbitrary at first. However, the diagram also imposes restrictions on the possible velocity laws for a fixed $\mathcal{S}$. For instance, in the simple example $\mathcal{S} = \mathbb{R}^3$, two different representations Q_t and Q'_t might differ by means of the orientation of the Euclidean coordinate axes, in other words by a rotation, which can be encoded as a linear and time-independent map $T_t = T : \mathbb{R}^3 \to \mathbb{R}^3$. In this case, one can infer the transformation rule

$$v'_t(TQ_t) = v'_t(Q'_t) = \frac{d}{dt}Q'_t = \frac{d}{dt}TQ_t = T\frac{d}{dt}Q_t = Tv_t(Q_t). \qquad (2.8)$$

In addition to the introduced degrees of freedom for one particular geometry, the many possible representations may come also with different geometries. This implies that the distance relations in any given initial configuration of matter points $[\Delta_0]_\sim$ in general do not allow to fix all the geometrical facts compatible with the entire history of relational change $\lambda \mapsto [\Delta_\lambda]_\sim$. For instance, it may be the case that the distance relations in a given initial configuration of matter points can be well described in Euclidean vector space $\mathcal{S} = \mathbb{R}^3$, but that the relations then develop in a way that does not allow for an efficient description based on Euclidean geometry.

Considering such a vast amount of degrees of freedom, one may wonder how to find a representation (2.4) of a motion (2.2) that enables us to conceive a sensible description of change by means of (2.6) at all. One needs a criterion to distinguish good from bad choices of representations such that, for each case at hand, a well-suited representation can be found. The obvious criterion is to strive for the simplest and most informative description of the history of relations $\lambda \mapsto [\Delta_\lambda]_\sim$. Recalling again the discussion of the diagram in Figure 2.1 according to which a change in representation is likely to imply a corresponding change in the velocity field, even though the change to be described is one and the same, the application of this criterion does not only depend on the representation alone but also on the corresponding velocity field. In this spirit, one may in the first place guess an embedding of the configuration of matter in a geometry and a representation of the dynamics in familiar terms and then check if the predictions we get from (2.6) fit the empirical data. If at some point our dynamical evolution diverges from the empirically observed one, it may be due to the fact that v_t in this representation cannot account accurately for the actual change. Then we have to adjust the representation of $\lambda \mapsto [\Delta_\lambda]_\sim$.

Likewise, it may the case that a description in this representation is theoretically possible, but involves an utterly complicated velocity field, or *vice versa* a very difficult representation but a trivial velocity field. The reason may again be an inconveniently chosen representation, which then requires further adaptation to find an optimal balance between representation and the corresponding velocity field. Hence, we first have to make a choice with respect to an initial configuration $[\Delta_0]_\simeq$. Then the strategy is to "fine tune" the choice of representation in order to achieve the simplest and most informative description of the history of relations $\lambda \mapsto [\Delta_\lambda]_\simeq$. Usually, the choice of a three-dimensional geometry as embedding for the initial configuration is a good starting point: less dimensions are not empirically adequate; more dimensions risk to bring in extra structure that will not be needed. Thus, in short, the geometry is just a matter of representation of the change in the configuration of distance relations.

The pertinent question then is how the velocity field can be fixed in such a way that specifying initial conditions at an arbitrary time and plugging them into a law that fits into equation (2.6) is sufficient to yield the motion of the matter points at *any* time as result. It is here that constant parameters—such as mass, charge, total energy, constants of nature, etc.—or parameters at an initial time—such as momenta, a wave function, fields, etc.—come into play. They are the means to achieve this result: they are such that specifying a value of them and inserting that value into a law of the type of (2.6) enables us to capture the change in a given configuration of matter points in a manner that is as simple and as informative as possible, obtaining as output the change in the configuration of matter points—that is, the velocity field—for *any* time. Let us call the parameters that allow us to achieve this result the *dynamical structure* of a physical theory.

There is no reason to privilege dynamical parameters that are attributed to the matter points individually. Furthermore, even if there are such parameters, these do not represent intrinsic properties of the matter points. Thus, in classical mechanics, mass is a parameter that is attributed to the particles individually, but it is not an intrinsic property, as argued at the beginning of section 2.1. In quantum mechanics, then, the parameter of mass is situated on the level of the wave function. Due to entanglement, the wave function usually applies only to a configuration of matter points. Furthermore, already velocity is in classical mechanics not a parameter that can be attributed to points taken individually. Consequently, our proposal is not hit by the campaign of Butterfield (2006a,b) against *pointillisme*: Butterfield's *pointillisme* both against dynamics and geometry is situated on the level of what we call the dynamical structure. Our proposal for ontology is committed to matter points, but these are holistically individuated by the distance relations in which they stand; they do not have intrinsic properties.

In any case, a fundamental physical theory is such that it defines a dynamical structure for the configuration of matter as a whole. Our proposal makes this procedure intelligible: if the ontology is one of matter points standing in distance relations, it is obvious that by changing any one distance relation one thereby changes in principle all the other ones. A matter point cannot come closer to or get farther away from another matter point without thereby also changing the distances that it bears to in principle all the other matter points in the configuration. Hence, if there are only distance relations among matter points, there is no problem to conceive interaction—that is, correlated motion of the matter points—and it is to be expected that this correlation affects in principle also distant matter points, without there being a problem of action at a distance or non-local interaction.

In brief, if there are only distance relations and if consequently any change affects in principle all the distance relations in a given configuration, correlated motion is to be expected without it making sense to call for something that transmits such "action at a distance" or non-local interaction. What we have here can appropriately be called dynamical holism. Furthermore, such a dynamical holism is not limited to the domain of fundamental physics. For instance, if one maintains that the meaning of a belief consists in its inferential relations to other beliefs (semantic holism, see, e.g., Sellars (1956) and Davidson (1995)), then changing any one belief affects in principle all the inferential relations in a given belief system (see Esfeld (2001) for a comparison between holism in philosophy of physics and philosophy of mind). Thus, although correlated motion is the bedrock in the physical realm, one can give an explanation in terms of unification (see Kitcher (1981)) by showing that such a holism obtains also in other domains such as semantics.

As regards a physical theory, the task hence is to specify a dynamical structure such that, for any configuration of matter points given as initial condition, the dynamical structure fixes how the universe would evolve if that configuration were the actual one. The dynamical structure thus goes beyond the actual configuration of matter: it fixes for any possible configuration of matter what the evolution of the universe would be like if that configuration were the actual one. It thereby supports counterfactual propositions. Furthermore, formulated in this way, there is a commitment to determinism built into the dynamical structure of a successful fundamental physical theory. However, there is nothing suspicious about determinism: dynamical parameters figuring in laws that fix all the change, given an initial configuration of matter points, are the simplest and most informative way to capture change. It may turn out that, as a matter of fact, such a physical theory cannot be achieved. But if there are dynamical parameters that fix only probabilities for how the configuration of matter points evolves, given an initial configuration, there always remains the question open whether one can do better—that is, find dynamical parameters that fix that change. To put it differently, once one has obtained such parameters, one knows that the work is done: one has achieved a description of the change that is both simple and maximally informative in requiring only one configuration of matter points as input in order to yield the whole past and future change as output. The question that remains then only is whether that description is empirically adequate and whether it can be further simplified without losing informational content. By way of consequence, as far as ontology is concerned, there is no reason to bring in probabilities. These come in when one ignores what the exact actual configuration of matter points is, as we will set out in the last section of Chapter 3.

2.3   The best of two worlds: Super-Humeanism about geometry and dynamics

The spatial structure is permutation invariant: as explained in section 2.1, although the distance relations individuate the matter points, labelling the matter points has no significance. The dynamical structure, by contrast, is in general not permutation invariant. It allows us to sort the matter points into different kinds of particles: some matter points move like charged particles, others like heavy or light particles, etc., so that they can be described as electrons, neutrons, etc. Therefore, a permutation of the particles' velocities obviously leads to a different physical situation. However, it would be unwarranted to conclude from this fact that some matter points are intrinsically electrons, others intrinsically neutrons, etc.

In any fundamental physical theory, the dynamical structure is defined by the configuration of the matter points as a whole: a law of the type of (2.6) specifies a velocity field for the whole configuration of matter in the universe, describing how that configuration changes by means of assigning a velocity to each matter point and thus singling out a curve $\lambda \mapsto [\Delta_\lambda]_\simeq$. This is particularly evident in the case of the quantum mechanical wave function whose evolution is defined on configuration space, with each point of that space representing a possible configuration of matter as a whole; due to entanglement, the wave function has, strictly speaking, to be assigned to the configuration of matter as a whole. However, even if a physical theory introduces dynamical parameters that are such that a value of them is attributed to the matter points taken individually, these parameters come in through their role in a dynamical structure that determines a velocity field for the configuration of matter as a whole. Thus, in the case of classical gravitation, for instance, the masses attributed to the particles are a constant that couples the motions of the particles to one another. In short, all the difference between the matter points originates in different ways in which they move, which can be described in an efficient manner by attributing values of various dynamical parameters—including constants of nature—to them.

By way of consequence, as in the case of the matter points individuated through the distances among them, so also when it comes to their dynamics, there is no need to admit properties in the sense of intrinsic physical properties making up for an intrinsic essence of the matter points. Structure is all there is—distance relations as far as the individuation of the fundamental objects is concerned, dynamical ones as far as the change in their configuration is concerned. We seek to capture the dynamical structure by conceiving various dynamical parameters, as we seek to capture the spatial structure by conceiving, for instance, an Euclidean space into which the matter points are embedded. However, as that space does not exist as an absolute space in the world, but is our means to represent the distances among the matter points, so the various dynamical parameters that physical theories introduce do not exist in the world: they are our means to represent the change in the distance relations among the matter points. As Hall (2009, § 5.2) puts it,

> The primary aim of physics—its first order business, as it were—is to account for *motions*, or more generally for change of spatial configurations of things over time. Put another way, there is one Fundamental Why-Question for physics: Why are things located where they are, when they are? In trying to answer this question, physics can of course introduce *new* physical magnitudes—and when it does, new why-questions will come with them. (So it is no part of the thesis we are considering that physics is *only* concerned with explaining motions; it is just that the other explanatory demands on it are, in a certain sense, derivative on this one.)

What Hall alludes to in this quotation is the crucial distinction between what we call the *primitive ontology* and the *dynamical structure* of a physical theory. The primitive ontology is the claim about what there simply is, whereas the dynamical structure includes all the parameters that enter into the theory in terms of their causal role for the evolution of what there simply is. Matter points being individuated by distance relations that change is our proposal for a primitive ontology that is independent of particular physical theories, being a proposal for a primitive ontology of the whole of physics. That proposal is minimally sufficient for empirical adequacy in a scientific realist spirit. It is backed up by the Cartesian argument that the commitment to extension is the only substantial commitment in an ontology of the natural realm, by the fact that positions of simple, discrete objects (particles) are the central and primitive parameter in classical as well as quantum physics and by the fact that all experimental evidence consists of the relative positions of discrete objects and their change (motion).

Nonetheless, as explained at the beginning of section 2.2, positions of discrete objects are not sufficient as parameter to obtain a simple law, because a given configuration of objects described in terms of relative distances does not provide any information about how the distances change. Hence, it cannot be sufficient as a basis for formulating a law of that change (at least a law that satisfies common standards of being simple and being informative). That is why further parameters have to be introduced in terms of their causal role for that change—in particular, parameters that are attributed to the matter points individually in a configuration that is given as initial condition (such as mass and charge in classical mechanics) or to the configuration as a whole (such as the wave function in quantum mechanics). That is to say, these parameters have to be specified as initial conditions in addition to the relative positions of the matter points in order for a law of the type of (2.6) to yield the evolution of the relative positions of the matter points (i.e. their motion).

Adding these parameters does as such not lead to an improvement in simplicity. Fixing values for them may turn out to be as complicated and messy as specifying initial relative positions of the matter points (consider the case of the quantum mechanical wave function). The gain in simplicity is that by figuring out these further parameters over and above the relative particle positions in an initial configuration, one determines an initial condition that can be inserted as input into a law of the type of (2.6) so that one obtains as output (in the deterministic case) a description of the whole past and future evolution of the particle configuration. It is therefore misleading to call these parameters nomological: they are not laws. They are part of the dynamical structure in that they are introduced in terms of their causal role in the evolution of the distance relations, but they are a part of the dynamical structure that has to be fixed as an initial condition in order to obtain a simple law that describes the whole evolution of the configuration (instead of a long list that merely enumerates how the configuration evolves) (cf. the distinction between two types of physical state in Bhogal and Perry (2017) —one type that consists only in the relative particle positions, and a further, broader type that includes what has to be specified as initial condition in addition to that).

However, their being part of the initial condition suggests *prima facie* to subscribe to an ontological commitment to these parameters along with the commitment to relative particle positions. In other words, at a first glance, they look like stuff that there is in the world together with the distances individuating the matter points. Consequently, the matter points appear as being equipped with masses and charges, as well as existing in fields, and be it a field like the quantum mechanical wave function. Nonetheless, the distinction between primitive ontology and dynamical structure points out that also for the scientific realist, there is no automatism from a parameter figuring in the dynamical structure of a well-established physical theory to subscribing to an ontological commitment to that parameter along with the elements of the primitive ontology. This issue has to be settled by philosophical argument. The benchmark for evaluating these candidates for an ontological commitment is whether including them in the ontology enhances the overall coherence of the position in that it provides for an understanding or an explanation of nature that is not available on the basis of a commitment to a minimalist primitive ontology only.

In this section, we will first frame our proposal in terms of what is known as Humeanism in contemporary metaphysics in order to show how laws enter into the picture on the basis of a commitment to primitive ontology only (the Humean mosaic). We will then argue that enriching the ontology by including those parameters that have to be specified as initial conditions in addition to the relative particle positions does not lead to a gain in explanation, but instead to new problems and serious drawbacks (thus decreasing instead of increasing the overall coherence of the position).

On Humeanism, what is referred to as the primitive ontology in contemporary philosophy of physics is the entire ontology in the sense that everything else supervenes on the primitive ontology: given the primitive ontology, everything else in the natural world is also given. Supervenience is here understood in the sense of reduction: everything reduces to the primitive ontology. Hence, whatever else there may be, it is no addition to being, it comes with the primitive ontology. In other words, the primitive ontology is sufficient to make true all the true propositions about the natural world (as mentioned in the introduction, we have no intention here to apply this reductionism to the mind, consciousness and normativity).

The way in which the change in the configuration of matter that is accessible to us occurs exhibits certain patterns or regularities. Conceptualizing these patterns or regularities, according to what is known as the best system account, the laws of nature are the theorems of the system that achieves the best balance between being simple and being informative in describing the evolution of the configuration of matter. A purely logical system would be very simple, but not informative at all. An extremely long inventory listing how the distances among the matter points change would yield the complete information about that change, but would not be simple at all. Laws of nature simplify and are informative at the same time, striking the best balance between these two virtues (see notably Lewis (1973b), pp. 72–75, Lewis (1994), section 3, Loewer (2007), Hall (2009) and Cohen and Callender (2009)).

Strictly speaking, what the laws of nature are is fixed only by the *entire* evolution of the configuration of matter—that is, the whole change in the distance relations among the matter points (cf. Beebee and Mele (2002), pp. 201–205). Of course, that whole change is not accessible to any observer in the universe. The axioms and theorems that figure in our fundamental physical theories are the best conjectures about the laws of nature that we can make by putting forward on the basis of the evidence that is available to us a system that achieves the best balance between being simple and being informative in accounting for that evidence. Consequently, what we regard as the laws of nature changes as more evidence becomes accessible to us.

Once laws are vindicated, Humeanism can handle causation in the same manner: causal relations may simply be given in terms of the regularities that the whole change in the configuration of matter exhibits (this was Hume's famous view see Hume (1739), book I, part III, and Hume (1748), section VII; see Baumgartner (2013) for a contemporary regularity theory of causation). Alternatively, the Humean can go for a sophisticated regularity theory in terms of counterfactuals (see Lewis (1973a, 2004)). The decisive issue then is that the truth-value of the counterfactuals expressing causal relations supervenes on the configuration of matter and its change in the actual world. To put it differently, no realism about other possible worlds is required to obtain truth-makers for counterfactuals in Humeanism (see Armstrong (2004), p. 445, and Loewer (2007), pp. 308–316).

In a famous passage, Lewis states the ontology of Humeanism in this way:

> Humean supervenience is named in honor of the greater denier of necessary connections [i.e. David Hume]. It is the doctrine that all there is to the world is a vast mosaic of local matters of particular fact, just one little thing and then another. (. . .) We have geometry: a system of external relations of spatio-temporal distance between points. Maybe points of spacetime itself, maybe point-sized bits of matter or aether or fields, maybe both. And at those points we have local qualities: perfectly natural intrinsic properties which need nothing bigger than a point at which to be instantiated. For short: we have an arrangement of qualities. And that is all. There is no difference without difference in the arrangement of qualities. All else supervenes on that.
>
> (Lewis (1986b), pp. ix–x)

Lewis describes the "vast mosaic of local matters of particular fact" not only in terms of primitive point objects but also in terms of local qualities in the sense of "perfectly natural intrinsic properties" instantiated by these point objects. This commitment puts Humeanism in a vulnerable position. Given that properties are introduced in physics through the role that they play for the change in the configuration of matter, that role being expressed in terms of parameters figuring in dynamical laws (cf. the quotation from Jackson (1998) at the beginning of section 2.1), the Humean who takes instantiations of natural properties to make up the local matters of particular fact has to maintain that this role is not essential to these properties. What role a given property plays depends on the laws of nature, which in turn depend on the *whole* evolution of the local matters of particular fact. Hence, what that role is cannot be a local affair. Furthermore, if it were essential for a given property to play a certain role, Humeanism would no longer eschew the commitment to necessary connections: the instantiation of such a role would impose modal, if not necessary connections on the local matters of particular fact. Thus, for instance, on Humeanism, the role that the property of mass plays for gravitational attraction in the world is contingent instead of essential to that property so that, in another possible world, mass can play a totally different role, depending on what the mosaic of local matters of particular fact is like in that other world. To illustrate this issue further, the properties of mass and charge can swap their roles: there is another world possible, w^*, in which the property that we pick out as charge in the actual world w plays the role of the property that we pick out as mass in the actual world w, and *vice versa*. Of course, w and w^* would be indiscernible: in w and w^* happen exactly the same things, matter moves in exactly the same way in both worlds. Nonetheless, w and w^* are two different worlds on Lewis's version of Humeanism.

Consequently, Lewis's Humeanism is committed to conceiving properties as pure qualities, known as quiddities. Furthermore, we cannot have epistemic access to these qualities. Our epistemic access is limited to the roles that the properties play for the evolution of the configuration of matter as described by a physical theory. Thus, we can distinguish between properties only insofar as they exercise different roles. There is no means to distinguish between what is supposed to be different pure qualities. The lack of epistemic access to properties conceived as pure qualities is known as humility. Lewis (2009) bites the bullet of endorsing humility and quidditism (that is, the commitment to properties as pure qualities). However, the rather baroque metaphysics of pure qualities with its implication of possible worlds differing only in the pure qualities that are instantiated in them is a heavy burden on Humeanism, in particular given that Humeanism sees itself as a metaphysics that is close to science and empiricism, avoiding any sort of occult metaphysics (see notably Black (2000)).

Fortunately, there is no need to take the Humean mosaic to be built up by instantiations of natural properties that then are pure qualities (quiddities). Instead, it is reasonable to conceive it in terms of a primitive ontology—namely, the spatial configuration of simple objects and its change. As shown here, such a primitive ontology is not committed to admitting space and time as substances, but can be formulated in terms of relationalism about space and time. Consequently, *the Humean mosaic consists only in matter points individuated by distance relations and the change of these relations.*

If that mosaic contained pure qualities, these qualities could not have any significance for the supervenience of laws on the mosaic. No purely qualitative variation could make up for a supervenience basis for laws. What accounts for the mosaic of local matters of particular fact being a supervenience basis for laws is that there is a variation in it that amounts to a discernible difference between matter points and change in that variation. For such a variation and change to obtain and for it to exhibit patterns or regularities such that laws can supervene on it it is sufficient that there are distance relations which individuate simple (= propertyless) objects and which change. Consequently, abandoning Lewis's natural properties does not result in any arbitrariness in defining the Humean mosaic: these natural properties do not have any significance for the supervenience basis anyway. What is needed are fundamental relations that individuate simple objects so that everything else can then supervene on this mosaic of relations and their change. As argued in section 2.1, distance relations individuating matter points perform that job when it comes to the natural world (in other words, they are—the only—natural relations). Hence, again, bringing in more—a commitment to mass, charge and the like as natural properties that are pure qualities in this case—creates new drawbacks (namely the commitments to quidditism and humility) instead of having any explanatory value.

This new version of Humeanism can be dubbed *Super-Humeanism*. That term is coined in analogy to the term "super-substantivalism" in the philosophy of space-time: while the standard substantivalist holds that space or space-time is a substance and that there is matter (particles, fields) filling space-time, the super-substantivalist maintains that there only is space-time, with what we take to be matter being reduced to the geometry of space-time. By the same token, while the standard Humean holds that there are spatial or spatio-temporal relations connecting points and natural intrinsic properties instantiated at these points, the Super-Humean maintains that there are only sparse points that then are matter points with distance relations individuating these points, but neither is there an underlying space nor are there natural intrinsic properties.[2]

Indeed, super-substantivalism can also be conceived as a Super-Humeanism: in this case, metrical relations individuate space-time points, with there being a plenum of space-time points. These points do not instantiate intrinsic properties; all there is to them is their being connected by metrical relations. As regards Lewis himself, in being committed to "perfectly natural intrinsic properties" instantiated at space-time points, he is also committed to space-time substantivalism. However, as he notes in Lewis (1986a, p. 76 note 55), a dualism of both substantival space-time and matter is uneconomical.

In fact, Lewis inclines towards resolving this dualism in terms of super-substantivalism. It is not clear from Lewis (1986a, p. 76 note 55) whether he is prepared to go beyond what Sklar (1974, pp. 166–167, 222–223) dismisses as a "linguistic trick"—namely, simply to attribute to space-time points the properties that are commonly ascribed to material objects (such as, e.g., mass and charge); such a move obviously cuts no ice, neither in physics nor in philosophy. The view that cuts at least philosophical ice is the project to reduce the properties commonly attributed to material objects to genuine properties of space-time points—namely, their geometrical ones, viz. the metrical relations between the space-time points. This can be done in the framework of Humeanism by interpreting the properties commonly attributed to material objects as being part and parcel of the best system, along with the laws, which then is the best system with respect to the mosaic of spatio-temporal relations between points.[3]

However, this is just what the Super-Humean does: in virtue of standing in distance relations, the points between which these relations hold are matter points, because they fulfill the Cartesian criterion for matter—namely, extension and motion. But these points then are sparse, and the relations cut down to distances that leave the space-time geometry as a matter of representation. The commitment to a plenum of space-time points (i.e. a substantival space-time) and fully-fledged spatio-temporal relations (i.e. coming with a certain geometry) then is unmotivated; it only results in a commitment to surplus structure, and it infringes upon the empirical adequacy of this view (cf. the remarks in section 2.1). Hence, from a Super-Humean perspective, super-substantivalism is cut down to a relationalism that only admits distance relations that individuate point objects, which then are matter points in virtue of their standing in these relations.

If one seeks for properties, they come into the picture of Super-Humeanism through the dynamical parameters that figure in the laws. Suppose, for the sake of the argument, that the laws of classical mechanics and classical electrodynamics are part of the best system. Mass, charge and other parameters figure in these laws. Both inertial and gravitational mass as well as charge are admitted in classical mechanics only because they perform a certain causal role as described by the laws of gravitation and electrodynamics—namely, to accelerate the particles in a certain manner. Hence, on Super-Humeanism, parameters or properties like mass and charge are no addition to being. The configuration of primitive objects and its change is all there is. Given that this change exhibits certain patterns, laws can be formulated, and given the parameters that figure in the laws and that are required, over and above positions, to specify an initial condition, one can attribute properties like mass and charge to the primitive objects–that is, the particles in classical mechanics. But the particles do not have these properties *per se*, as something essential or intrinsic to them. They obtain them only through the regularities that the change in the distance relations among them exhibits via the attempt to capture these regularities in simple laws (see Hall (2009), § 5.2; see also Martens (2017)). That is to say: it is not mass and charge qua properties belonging to individual matter points that determine their trajectories by means of the causal role that they play in the laws of classical mechanics and electrodynamics; on the contrary, the trajectories that the matter points take throughout the evolution of the universe make it that parameters such as mass and charge figure in the dynamical laws such that a value of mass and charge applies to the matter points taken individually and is used to define the initial conditions that allow the application of the laws.

It is therefore more appropriate to set this ontology out in terms of predicates and truth-makers for predicates. The distance relations among the matter points and the change in these relations make true all the true propositions about the world, including in particular the propositions expressing laws of nature. Hence, if the laws of classical mechanics figure in the best system, predicates such as "mass" and "charge" apply to the particles in virtue of the patterns that the particle trajectories exhibit. These predicates as well as all the other ones appearing in the propositions that are true about the world really apply, and the propositions really are true, there is nothing fictitious about them. But what there is and hence what makes them true is nothing over and above matter points individuated by distance relations and the change in these relations.[4]

In brief, on Super-Humeanism, not only the laws but also the dynamical parameters that a physical theory introduces as well as the geometry of space and time (see Huggett (2006) on geometry and inertial frames in classical mechanics) supervene on the evolution of the configuration of matter in the universe, which is then characterized only in terms of the distance relations connecting propertyless matter points. Super-Humeanism can therefore be conceived as being inspired by the following quotation from Russell (1927):

> There are many possible ways of turning some things hitherto regarded as "real" into mere laws concerning the other things. Obviously there must be a limit to this process, or else all the things in the world will merely be each other's washing.
>
> (Russell (1927), p. 325)

Thus, mass, charge, the geometry of space and time, etc., are "mere laws concerning the other things". The bedrock—the Humean mosaic—is matter points individuated by distance relations that change. All the rest are parameters introduced to capture that change in a manner that is both simple and informative about that change, given that all the experimental evidence consists in that change. Super-Humeanism thus is a simple and parsimonious ontology that is close to science, being free of the burden of the baroque metaphysics of quiddistic intrinsic properties as in Lewis's Humeanism.

By the same token, turning to quantum physics with particles as the primitive ontology (as in Bohmian mechanics), on Super-Humeanism, it is not the wave function that determines the trajectories of the particles; on the contrary, given the motion of the particles, the regularities that their motion exhibits make it that a wave function parameter figures in the dynamical laws capturing that motion and has to be specified as initial condition in order to apply the laws. Miller (2014), Esfeld (2014b), Callender (2015) and Bhogal and Perry (2017) have worked this stance out independently of one another as regards the wave function in quantum theories with a primitive ontology (such as Bohmian mechanics) (see Dickson (2000) for a forerunner). There is a precise physical model in this sense by Dowker and Herbauts (2005), namely how to conjecture the wave function from a history of GRW flashes on a lattice. In this model, it is clear how the wave function simplifies by figuring in a law for the evolution of the GRW flashes. However, if it turned out that the universal wave function were as complicated and messy as the entire history of the relative particle positions (or the flashes), then the wave function would be inappropriate as a parameter that enters into a law that describes that history; there would then be no gain in simplicity: the whole information about the physical world is trivially contained in the entire history of the particle positions.[5]

On Super-Humeanism, in brief, there are no relations of entanglement in nature over and above the distance relations among the matter points. The quantum state does not mysteriously tie the motion of the Bohmian particles together by somehow guiding or piloting them in a coordinated manner. There is no quantum state in nature, although all the propositions formulated in terms of the quantum state are perfectly true. As a contingent matter of fact about the actual universe, the way in which the particles move is such that it manifests certain stable correlations so that, if we set out to represent their motion in a manner that is both simple and informative, we have to write down an entangled quantum state and a law in which an entangled wave function figures. Given that law, we then have a truth-maker for the counterfactual propositions about what would happen to the motion of other particles if one intervened on the motion of a specific particle in a certain manner, etc. But there are no mysterious influences going from one particle to distant particles via a quantum state in nature.

That notwithstanding, Super-Humeanism is distinct from instrumentalism about the wave function: the wave function is the central dynamical parameter, figuring in the law of motion for the particles, as well as being itself subject to an evolution according to a law (i.e. the Schrödinger equation). These laws then are linked, as we shall explain in Chapter 3, sections 2 and 4, with the probability calculus in which the wave function is employed to calculate probabilities for measurement outcomes according to Born's rule. In brief, the first and foremost role of the wave function is the one that it has in the law of motion for the primitive ontology instead of being an instrument to calculate probabilities. In a nutshell, Super-Humeanism allows us to have the best of two worlds: simplicity in ontology achieved through parsimony and simplicity in description achieved through buying into the simplest physical theory that is empirically adequate. Super-Humeanism thereby is a form of scientific realism.

Last but not least, Humeanism makes clear that there is nothing suspicious about a physical theory being deterministic: there is nothing that determines the evolution of the configuration of matter. That configuration develops in a certain manner as a contingent matter of fact, exhibiting certain stable patterns or regularities, which serve as truth-makers for laws of nature. But this does not mean that the laws or any of the parameters figuring in them have the power to determine the evolution of the configuration of matter, or to make that evolution necessary. Considering any one given state of the configuration of matter points consisting in certain distance relations among the matter points, there is nothing in that state that poses a constraint on how the distance relations can evolve (see notably Beebee (2000), Beebee and Mele (2002), Beebee (2006) and Hall (2009), § 0). Humeanism thus makes intelligible why physics strives for deterministic theories, because these are the foremost candidates for theories that strike the best balance between being simple and being informative. This is so even if it should turn out that for any deterministic theory in physics—such as, e.g., Newtonian mechanics—initial conditions are conceivable in which the determinism breaks down. Humeanism can easily accommodate such marginal failures of determinism, since it does not consider determinism as a modal feature of the world, but only as a means to achieve an optimal balance between simplicity and strength in a system that represents the evolution of the particular matters of fact that make up the universe.

A fortiori, it is a misunderstanding to think that determinism in the dynamical structure of a physical theory has implications for human free will. The issue of such implications may arise only if one conceives the dynamical structure as an additional element of the ontology—that is to say, it belongs to the ontology in addition to the spatial structure, being a further primitive over and above the matter points individuated by the distance relations in which they stand and their change, such that, loosely speaking, the dynamical structure pushes the configuration of matter in a certain direction. By the way, if this were so, one realizes immediately that the ensuing conflict with human free will depends in no way on the dynamical structure being deterministic or stochastic: the tension with free will is the same whether the physical laws make the configuration of matter (including the matter of human brains) move in a certain direction with a certain probability that does not depend on human free will or whether the physical laws do so in a deterministic manner (see Loewer (1996)). Again, reifying the dynamical structure entails new problems.

However, the Super-Humean ontology provokes the objection that by being most parsimonious, it loses explanatory value. We already considered and rebutted in section 2.1 the claim of Maudlin (2007a, pp. 87–89) according to which the substantivalist about space can explain the features of the distance relation that the relationalist has to accept as primitive. In brief, there is no gain of explanation in substantivalism, because the features at issue come in a package with the postulation of an absolute space. In general, where our minimalist ontology admits primitive distance relations and primitive change of these relations that exhibits patterns such that the relations and their change can be represented, for instance, in terms of Euclidean geometry and masses, charges, forces, fields, wave functions, etc., as they figure in the dynamical equations of a physical theory, a richer ontology has an underlying space filled not only with point particles but also with mass and charge as intrinsic properties of these particles as well as forces, fields, or wave functions, etc.

The objection then is that thus enriching the ontology does not result in an explanatory gain: these additional entities are dynamical parameters that are *defined* through the functional or causal role that they play for the evolution of the particle configuration. For instance, one does not give a deeper explanation of attractive particle motion by admitting mass or the gravitational force to the ontology, because these parameters are defined through the effect that they have (or can have or are the power to have) on the motion of the particles. The only difference then is that what is contingent on the minimalist ontology comes out as necessary on the richer ontology (the Euclidean geometry is necessary given an underlying Euclidean space, the attractive motion of the particles is necessary given masses and the gravitational force, etc.). However, lifting the status of something from contingent (because accepted as primitive matter of fact) to necessary does not yield a deeper explanation. Taking explanations to end in the distance relations among matter points and their change endows our minimalist ontology with all the explanatory value that one can reasonably demand, namely to explain all the other phenomena in terms of the fundamental physical entities.

The most prominent way to enrich the ontology is to admit properties that are dispositions or powers—that is, the disposition or power to change the motion of matter in a certain manner under certain conditions. By contrast to Lewis's conception of natural properties that are pure qualities, properties conceived as dispositions or powers establish modal connections in nature in the sense that it is necessary that a certain change in the motion of the matter points occurs if such properties and conditions are present. In contrast to Humeanism, the laws of nature then supervene on these properties in the sense that the laws express which dispositions or powers these properties are (see notably Bird (2007)). Consequently, what the laws of nature are does not depend on the evolution of the configuration of matter. Quite to the contrary, the manner in which this evolution occurs is brought about by these properties. Hence, if all the dynamical parameters are instantiated by the configuration of matter at any given time, what the laws of nature are is fixed by the properties that the configuration of matter has at any arbitrarily chosen time.

However, by admitting mass and charge as intrinsic properties of the matter points over and above the distance relations among them, one does not provide an explanation of the change in the distance relations that is illuminating by contrast to accepting that change as a primitive matter of fact as the Humean does. It is true that one traces the change that occurs back to properties of the matter points. But these properties are defined in terms of the causal role that they exert for the motion of the matter points. Thus, one shifts what has to be accepted as primitive from the change that occurs to the causal role of certain properties. This is no gain in explanation. It is an instance of the scheme of which Molière makes a caricature in *Le malade imaginaire*: one does not explain why people fall asleep after the consumption of opium by attributing a dormitive power to opium—although, of course, mass and charge are sparse, fundamental properties by contrast to the phenomenological properties of opium. Nonetheless, the Molière argument hits also these latter properties: like the dormitive power of opium, they are defined in terms of the effects that they bring about under certain conditions.

This situation does not change if one draws attention to the fact that parameters such as mass and charge figure in dynamical laws that are connected with the symmetries of space-time. The same goes for the position that admits not only the space-time symmetries as primitive, coming with the ontological commitment to a substantival space-time, but also the dynamical laws as primitive (see notably Maudlin (2007a)). One can then obtain beautiful explanations in terms of unification. But insofar as scientific explanations are concerned, the Humean is entitled to vindicate them in the same way as the opponent who endorses an ontological commitment to space-time symmetries and modal entities in nature (dispositions, powers, primitive laws). For the Humean, space-time geometry and the laws including the parameters that figure in them come in as the package that provides the most simple and most informative description of the configuration of matter and its change, thus yielding scientific explanations through unification (for the recent debate about explanations in Humeanism, see notably Loewer (2012), Lange (2013), Miller (2015) and Marshall (2015)).

Admitting mass and charge as intrinsic properties of the matter points that are dispositions or powers again introduces a surplus structure in the ontology: the dispositions or powers are not always manifest, and there may be dispositions or powers existing in the universe without the conditions for their manifestation ever being fulfilled (the same goes for primitive laws: they may not always be instantiated). Moreover, again, admitting dispositions or powers brings in new drawbacks instead of adding explanatory value. As reifying space leads into the impasse of having to answer the question of what distinguishes matter from space elaborated on in section 2.1, so upgrading the dynamical parameters that a physical theory introduces to intrinsic properties of the physical objects leads into the impasse of having to answer the question of how a physical object can reach out to other objects and change their motion in virtue of properties that are intrinsic to it—since the effect that a particle produces in virtue of instantiating properties such as mass and charge is a change in the motion of *other* particles.

The search for an answer to this question then often results in the view of there being forces or fields that literally propagate in space by means of which one particle changes the motion of other particles. Such views, however, have been convincingly criticized as anthropomorphism by Russell (1912) and others, and their complete breakdown is evident at the latest when it comes to quantum non-locality: the correlated motion of quantum particles cannot be explained in terms of a force or field propagating in space, and the wave function is defined on configuration space instead of being a field in physical space. In general, the ontological status of such mediating entities remains unclear: Are forces or fields some sort of stuff filling space in addition to the particles? Are they properties of something? But of what? Furthermore, they neither are dispositions nor manifestations of dispositions: the manifestation of mass and charge qua dispositions of particles is the change in the motion of other particles, not the force or the field, but the force or the field are not dispositions in their own right either, since they depend on the mass and charge qua dispositions of the particles (we will go into fields in classical electrodynamics in detail in section 5.2).

One can avoid these additional drawbacks by switching from intrinsic properties to structures. That is to say, instead of admitting intrinsic properties that are dispositions or powers, one endorses dynamical relations among the matter points that exist over and above the distance relations and that are the disposition or power to change the distance relations in a certain manner—that is, the disposition or power for a certain correlated motion of the matter points. Thus, one maintains that the dynamical structure of a physical theory refers to a dynamical structure that exists in the world and that is a modal structure, by contrast to the spatial structure. The most prominent example of this strategy is admitting entanglement relations among quantum particles as relations in which these particles stand over and above the distance relations, with these relations being the disposition or the power to manifest themselves in a certain correlated motion of the quantum particles (as, for instance, in the correlated outcomes of the EPR experiment). But one can also conceive the dynamical structure of classical mechanics as referring by means of mass and charge, conceived now as coupling parameters, as well as forces and fields to dynamical relations in the world that are dispositions or powers and that manifest themselves in attractive and repulsive motion of the particles.

Indeed, it is common among the proponents of ontic structural realism to regard the dynamical structures as irreducibly modal structures (see notably Esfeld (2009) and French (2014), chs. 9 and 10). The resulting view then is a structuralist conception of the laws of nature in distinction to the Humean one: the laws are our manner to capture the dynamical structure that there is in nature (cf. Cei and French (2014)). That structure is irreducibly modal in that it constrains the evolution of the distance relations among the matter points. Note, as already mentioned in section 2.1, that it is a misconception to enquire whether these structures can individuate the objects that stand in them and, in case they fail to do so, to conclude that there is a problem with the basic physical objects being conceived as individuals: the structure of entanglement, for instance, is encoded in the quantum state, which is defined on configuration space—that is, the mathematical space each point of which represents a possible particle configuration in physical space. Consequently, the entanglement structure *presupposes* a particle configuration to which it is applied and cannot be that what individuates these particles.

However, as its counterpart in terms of admitting modal, intrinsic properties of the basic objects, so admitting modal relations in which these objects stand over and above the distance relations does not provide additional explanatory value, since this move is also subject to a Molière-type objection: the modal structures are defined in terms of what they do —or can do or are the power to do—for the particle motions. For instance, one does not give a deeper explanation of quantum non-locality in terms of a modal structure of quantum entanglement, because this structure is defined in terms of correlating the possible ways in which the quantum objects that instantiate this structure can evolve.

It is true that the commitment to modal, dynamical relations instead of modal, dynamical intrinsic properties does not create the drawback of how one particle can reach out to changing the motion of other particles in virtue of properties that are intrinsic to it, since the particles are *ab initio* dynamically related on the modal structural realist view. However, this view faces another problem—namely, to vindicate the dynamical structure as a network of concrete physical relations among the particles by contrast to an abstract mathematical structure that represents the motions of the particles in a simple and informative manner (as Humeanism maintains). Prominent proponents of ontic structural realism simply refuse to acknowledge that problem, being prepared to concede that it is not possible to distinguish concrete physical from abstract mathematical structures (see, e.g., Ladyman and Ross (2007), pp. 159–161, and French (2014), p. 230), but if this were so, it would be a fatal blow to ontic structural realism as a proposal for the ontology of physics by contrast to the ontology of mathematics (see, e.g., Briceno and Mumford (2016) for a recent criticism in that sense).

It is also not sufficient to maintain, as one of us did in earlier publications (Esfeld (2009)), that, for instance, the relations of quantum entanglement are concrete physical relations by contrast to abstract mathematical ones, because they are causal in the sense that they have an effect on the motion of the particles (or are the disposition or the power to have such an effect), since what is at issue precisely is whether there is anything that has such an effect qua being a concrete physical relation. The distance relations set the paradigm for concrete physical relations: they individuate the matter points. As brought out most clearly by quantum entanglement, if there is a dynamical relation of quantum entanglement over and above the distance relations, putting a constraint on how these relations can change, there is only one instantiation of such a relation that encompasses all the matter points as its relata, since there is only one, non-separable quantum state of the whole configuration of matter (as represented by a point in the configuration space of the universe with the wave function acting on that point).

In other words, the universal wave function in quantum mechanics is entangled in the sense that it binds the temporal development of each particle to the temporal development of strictly speaking all the other particles. This fact brings out clearly the problem how to vindicate that dynamical structure as a network of concrete physical relations by contrast to an abstract mathematical structure that is part and parcel of the means to represent the real physical relations and their evolution—namely, the distance relations and their change—in a simple and informative manner. In a nutshell, the objection is that by committing itself to a modal, dynamical structure existing in nature, ontic structural realism reifies what is the mathematical representation of the evolution of the physical relations to a mysterious structure behind the scene of the changing distance relations that binds the motions of the matter points together. Again, this is evidence of the general claim made in this book: bringing in more than matter points individuated by distance relations and change in these relations creates new drawbacks instead of providing additional explanatory value.

However, our concern here is not the metaphysics of modality. Humeanism enters the proposal sketched out here only as a strategy to combine simplicity in ontology with simplicity in representation—in other words, as a strategy to maintain scientific realism without building ontological commitments on the representational means that physical theories employ. The proposed minimalist ontology is neutral about the metaphysics of modality. Due to the axiom that the distance relations between the matter points change, one may in fact receive this ontology as implying that it is in the nature of the distance relations to change, or in the nature of the matter points to change the distances among them. Such a formulation leaves open whether there is anything in the nature of the matter points or the distance relations that makes certain ways of their change necessary (cf. the interpretation of Strawson (1989) of the historical Hume according to which Hume denies only our epistemic access to necessary connections, but not their very existence). In other words, we remain agnostic about whether there is a reason (*logos*) driving the evolution of the *cosmos*. Thus, the answer to the question why we are so successful in representing the change that is accessible to us in terms of maximally simple laws that are maximally informative about that change may be that there is a reason in the way in which change occurs, but the answer may also be that since change is eternal, every possible sequence of change in the distance relations among the permanent matter points actually occurs, and beings that are able to reason about that change can only exist in a sequence of change (say from a big bang to a big crunch) that is highly regular so that it enables a representation in terms of maximally simple and maximally informative dynamical equations.

The only claim that we make is that it is misguided to take the geometry and the dynamical parameters that enter into the laws of *our* physical theories for anything more than *our* attempts to achieve a description of the change in the distances among the matter points that is both simple and informative about that change; reifying the geometry and the dynamical parameters leads to artificial problems instead of satisfying the demand for a deeper explanation. That is why the ontology of the natural world is fully specified by the minimalism expressed in the two axioms of there being matter points individuated by distance relations and change of these relations.

Notes

1. The debate about weak discernibility goes back to Saunders (2006). As regards the distance relations, see the exchange between Wüthrich (2009) and Muller (2011). On the shortcomings of weak discernibility, see, notably, Dieks and Versteegh (2008, p. 926), Ladyman and Bigaj (2010, p. 130) and, for a recent overview and argument, Bigaj (2015).
2. Castañeda (1980, p. 106) uses the term "super-Humean world", meaning a view that does not regard energy (or forces) as something that exists in the world, but there is no rejection of absolute space or natural intrinsic properties considered in Castañeda. We are grateful to Gordon Belot for suggesting the term "Super-Humeanism" for our view of a relationalism that rejects intrinsic properties of the spatial relata.
3. In this case, one cuts philosophical ice. To cut physical ice, one would have to show how to do physics in terms of the geometry of space-time only, in spite of the failure of Wheeler's programme of geometrodynamics in the 1960s mentioned in section 2.1.
4. LeBihan (2016) proposes a super-relationalism that eliminates the objects standing in the distance relations, but keeps natural properties. The problem with this view is that all the candidates for natural properties from physics (such as mass, charge, etc.) are defined in terms of their function for the motion of physical objects such as point particles—that is, their function for the evolution of the distance relations among primitive objects that can be considered as being individuated by these very relations. Again, properties can be dispensed with in a parsimonious ontology, but not objects that stand in the basic relations.
5. Furthermore, there are attempts to formulate quantum mechanics without the need of wave functions at all, which are often termed many-interacting-worlds approaches; see Schiff and Poirier (2012), Hall et al. (2014) and Sebens (2015). The basic idea in these papers is to use continuous or discrete and finite collections of Bohmian trajectories to sample the wave function by means of their distribution and their velocities so that the wave function itself can be discarded in the dynamics. These trajectories may be interpreted as many coexisting classical worlds that on top of their inherent classical interaction are subject to a quantum interaction arising from the Bohmian potential, and quantum expectation values can be computed by means of Laplacian averages.

3 Minimalist ontology and dynamical structure in classical and quantum mechanics

3.1 From ontology to physics: two strategies for classical mechanics

Let us take up the representation of the distance relations between matter points in terms of vectors in Euclidean space that we have introduced in general terms in Chapter 2, section 2. This representation allows us to specify a law that records the change of the distance relations in terms of usual derivatives:

$$v_t(Q_t) := \frac{d}{dt} Q_t, \quad \text{for} \quad t \mapsto Q_t. \tag{3.1}$$

This law defines a vector field v_t, called the "velocity field", such that all possible motions $t \mapsto Q_t$ are integral curves of v_t.

As explained in Chapter 2, the task of physics then is to find a general law of motion that captures the change in the distances among the matter points. Hence, one can turn the definition in (3.1) around and study all motions $t \mapsto Q_t$ that fulfill

$$\frac{d}{dt} Q_t = v_t(Q_t). \tag{3.2}$$

For a configuration of matter points represented in Euclidean space $\mathcal{S} = \mathbb{R}^3$, using the same notation as in Chapter 2, section 2, to denote the motion $\lambda \to [\Delta_\lambda]_\sim$ by $t \mapsto Q_t = (q_{1,t}, \ldots, q_{N,t})$, one may set up a theory of classical, pre-relativistic gravitation as follows, working with a representation of the matter points as being inserted into an absolute Euclidean background space and as evolving in an absolute background time, as in Newtonian mechanics. The velocity field in (3.2) given as $v_t = (v_{1,t}, \ldots, v_{N,t}) \in \mathbb{R}^{3N}$ is ruled by the equations

$$\frac{d}{dt} v_{k,t}(Q) = -\frac{1}{M_k} \nabla_k V(Q), \quad \text{for all } k = 1, \ldots, N, \text{ and } Q \in \mathbb{R}^{3N}, \tag{3.3}$$

where the map $V : \mathbb{R}^{3N} \to \mathbb{R}$ is given explicitly by

$$V(Q) := -\frac{1}{2} \sum_{k=1}^{N} \sum_{j \neq k} G \frac{m_k m_j}{|q_k - q_j|}, \tag{3.4}$$

and $G, M_1, m_1, ..., M_N, m_N \in \mathbb{R}^+$ are additional parameters. Recall that $|q_{k,t} - q_{j,t}|$ equals $\delta(\Delta_{kj,\tau(t)})$ defined in Chapter 2, section 2. Together with further parameters denoted by $\dot{q}_{k,0} \in \mathbb{R}^3$, the equations in (3.3) uniquely determine a velocity field that is given by

$$v_t(Q_t) = (v_{1,t}(Q_t), ..., v_{N,t}(Q_t)), \quad \text{for}$$

$$v_{k,t}(Q_t) := \dot{q}_{k,0} + \int_0^t \left(-\frac{1}{M_k} \nabla_k V(Q_s) \right) ds. \tag{3.5}$$

These additional parameters—that is, the relationships described in (3.3) and (3.4), Newton's constant of gravitation G, inertial and gravitational masses M_k and m_k, and the initial velocities $\dot{q}_{k,0}$—make up the dynamical structure of this version of classical gravitation. They are the only degrees of freedom left in the choice of an admissible velocity field v_t. Thanks to (3.2), specifying these parameters together with the initial configuration $Q_0 \in \mathbb{R}^{3(N-1)}$ determines the motion $t \mapsto Q_t$ uniquely.

Huggett (2006) shows how one can understand Euclidean geometry and Newtonian mechanics in a package as the Humean best system for a world of classical mechanics that consists only in distances among point particles and their change.[1] As explained in Chapter 2, section 3, the idea is that if one considers the evolution of the distance relations in the configuration of matter points of the universe, the spatio-temporal geometry best suited to describe the universe is fixed together with the dynamical laws by the evolution that these relations take throughout the history of the universe. Given the fact that the change in the distance relations manifests certain salient patterns, Huggett (2006) singles out the notion of inertial motion as the idea of a particularly regular and simple motion. He then defines the notion of *adapted* reference frame as an assignment of real numbers at a given time—that is, a set of *d*-tuples such that (i) the origin of the frame—the $(0, 0, ..., 0)$ tuple—corresponds to the "position" at time zero of the matter point to which the frame is adapted and (ii) the distances along the *d* axes correspond to the distances from that matter point. The choice of *d* (the dimensionality of space) and of the distance—that is, the labeling of spatial relations (see definition 1 in Chapter 2, section 1)—is totally arbitrary. The only constraint is that this choice enables the best physical description: in the classical case, the best choice is $d = 3$ with the distance given by the usual Euclidean formula.

An adapted frame is a *bona fide* relational concept in the sense that it does not depend on any absolute concept, such as that of position in absolute space. In particular, it is well-suited for a Leibnizian-Humean stance. Given the fact that its definition relies only on the notion of matter point and arbitrary numerical assignments, an adapted frame can be easily constructed within our minimalist ontology. In fact, when we say that, for example, in a frame adapted to a particle i the simplest and most informative assignment that individuates another particle j at time t is a triple (x, y, z), we rely only on an arbitrary labeling of the instantaneous configuration the reference particle i is in (which is a well-defined concept given our Leibnizian view of time as a labeling for the succession of states of the universal configuration of matter points given in terms of the relative distances among the matter points), as well as on an arbitrary labeling of the distance relation between i and j. In short, an adapted frame is a mere representational means that requires nothing more than the relationalist Humean mosaic.

However, relying on adapted frames *only* might not be enough for reconstructing classical mechanics. For example, it might be the case that no matter point can be adapted to an inertial reference frame, thus blocking Huggett's strategy. In order to obviate this problem, unoccupied frames can be relationally defined from adapted ones by means of continuous rigid spatial translations. Roughly, a spatial translation from a frame O into another frame O′ amounts to shifting the set of d-tuples constituting O by a certain factor (defined, for instance, by a suitable continuous function that takes a d-tuple in O as input and gives back a d-tuple in O′), such that the relative distances between particles remain unaltered in the new frame. Also in this case, no ontological commitment going beyond those included in axioms 1 and 2 is called for: a spatial rigid translation so defined, in fact, relies only on numerical assignments and, hence, does not require any pre-existing geometrical notion such as that of affine connection.

Obviously, the notion of rigid spatial translation makes it possible to relate any kind of frame, independently of whether they are occupied or not. With this machinery in place, an inertial frame can then be defined as the frame in which the dynamical laws that supervene on the history of relations for the entire universe hold. In this way, the notion of inertial motion neither presupposes a substantival affine structure that singles out straight trajectories nor an absolute external time to which the uniformity of motion should be referred. Rather, these two structures supervene on purely relational facts. By the same token, absolute acceleration is reduced to the history of change of the spatial relations holding between an inertial and a non-inertial frame. Similarly, the regularities in the history of relations make it that Euclidean geometry is the simplest and most informative geometry representing that history. Such a framework clearly vindicates Leibnizian relationalism about space (spatial relations and their change are the ontological bedrock) and time (temporal facts supervene on the history—that is, an ordered sequence—of instantaneous distance relations).

In the same vein, as already mentioned in Chapter 2, section 3, Hall (2009, § 5.2) sketches out how dynamical parameters such as both inertial and gravitational mass as well as charge can be introduced as variables that figure in the laws of classical mechanics achieving the simplest and most informative description of the change in the relative particle positions throughout the history of the universe. Thus, the applicability of Super-Humeanism to Newtonian mechanics confirms that there is no point in reading off one's ontological commitments from the dynamical structure of physical theories: one can be a scientific realist with respect to Newtonian mechanics and yet be committed only to distance relations individuating matter points and the change of these relations in spite of the fact that Newtonian mechanics employs an absolute Euclidean background space and an absolute background time as well as various further dynamical parameters to represent that change.

To mention a particularly striking example of this fact, consider a model of Newtonian mechanics with an angular momentum of the universe J that is greater than zero in the centre-of-mass rest frame. Obviously, a rotating universe is not conceivable in a relationalist ontology that admits only distance relations among matter points, but no space in which the configuration of matter of the universe is embedded. However, also in a Newtonian ontology of a universe rotating in absolute space, the rotation of the universe would manifest itself in certain changes in the distance relations among the point particles (e.g. in inhomogeneities in the cosmic microwave background with respect to our point of view). Consequently, the relationalist is free to interpret a value of angular momentum of the universe that is greater than zero as a convenient means to capture those changes in a simple and informative manner, without being committed to a space in which the universe rotates: describing those changes by using only variables for the change of relative distances would lead to a law of motion that is extremely complicated. Introducing further variables—such as an angular momentum of the universe that can have a value greater than zero—by contrast, would enable the formulation of a law of motion that is simple and elegant, but that is there only to capture the change in the relative distances among the point particles. In this manner, the relationalist can handle all kinds of bucket-like challenges.

Nonetheless, this strategy cannot recognize all the possible mathematical solutions of the dynamical equations of a physical theory as describing physically possible situations. Axiom 1 and requirements (i) to (iv) of the minimalist ontology set out in Chapter 2, section 1, pose also a constraint on the dynamics: for instance, an evolution of the distance relations among the matter points that ends up in an entirely symmetrical configuration of the matter points of the entire universe is excluded in the same way as is a symmetrical initial configuration of the entire universe. Such solutions are a mathematical surplus of the formalism; the corresponding points in configuration space do not represent physically possible configurations of matter points of the entire universe. As argued in in Chapter 2, section 1, this is no objectionable restriction: having empirical adequacy in mind, there is no need to admit, for instance, entirely symmetrical worlds as physically possible worlds. As also argued there, this refusal by no means diminishes the importance of symmetries to obtain a simple and informative representation of the universal configuration of matter and its evolution.

The *Super-Humean strategy* is a purely philosophical one: In a nutshell, on this strategy, *there is a relationalist ontology, but a non-relationalist physical theory*. This is no problem, since the non-relationalist theory can be interpreted in a cogent manner that is consistent with scientific realism as being committed to no more than the relationalist ontology.

However, one may wonder whether buying into all the formal apparatus of, say, Newtonian mechanics is necessary to achieve a description of the change in the distance relations that is both simple and informative. This reflection opens up the way for another strategy to link the minimalist ontology up with physics that can be dubbed *alternative theory strategy*: instead of endorsing physical theories as they stand—such as Newtonian mechanics—and refusing to take their dynamical structure as guide to the ontology, one constructs alternative physical theories whose formal apparatus stays as close as possible to the ontology of there being only distance relations among point particles and their change and that matches the standard theories in their testable predictions. In a nutshell, this strategy consists in *building a relationalist physical theory on a relationalist ontology*.

It is important to be clear about what the alternative theory strategy can and what it cannot achieve: even if the ontology is exhausted by distance relations individuating matter points and the change of these relations, when it comes to formulating a law capturing that change, further dynamical parameters have to be introduced, since there is nothing about the distance relations making up any given configuration of matter points that contains information about the—past and future—change of these relations. Consequently, the alternative theory strategy has to admit dynamical parameters such as mass, constants of nature, initial momenta, etc., and cannot but resort to the Super-Humean strategy in order to ban these parameters from the ontology. However, when it comes to space and time, the aim of this strategy is to avoid quantities that are tied to absolute space and time, such as empty space-time points, absolute velocities, absolute accelerations, or absolute rotations. As regards classical mechanics, the alternative theory strategy goes back at least to Mach (1919). In the last three decades, it has been worked out by Barbour and collaborators (see Barbour and Bertotti (1982) for the seminal paper that laid down the "best-matching" framework).

Furthermore, Belot (1999) and Saunders (2013) have also proposed each a relationalist theory of classical mechanics.

Let us discuss Belot's proposal first (see also Pooley and Brown (2002), section 5, for an appraisal of this proposal). Belot starts by considering the Hamiltonian formulation of classical mechanics given in the language of symplectic geometry. For this we take Q to be the configuration space. The tangent bundle TQ is the collection of tuples $(q, \dot{q})$, where $q \in Q$ and $\dot{q}$ is a tangent vector (velocity vector) of a curve through q. The cotangent bundle T^*Q is, strictly speaking, the dual bundle of TQ (momenta are described as linear functionals of velocity vectors), which is the mathematical formulation of the so-called *phase space*. For simplicity (since both spaces are isomorphic), we will write $(q, \dot{q})$ for the elements of T^*Q, where $\dot{q}$ is to be understood as the corresponding momentum (see Frankel (1997), section 2.3c, for technical details). Phase space comes equipped with a smooth function H, called the Hamiltonian, which—roughly—represents the total energy of the system, and a two-form ω, called the symplectic form. Glossing over the technical aspects, ω renders it possible to define a map $H \mapsto X_H$ that associates to the Hamiltonian a smooth vector field X_H over T^*Q: such a map is nothing but an intrinsic representation of the usual Hamilton's equations. Hence, by integrating X_H given some initial conditions $(\mathbf{q}_0, \dot{\mathbf{q}}_0)$, we get the unique curve in T^*Q that represents the dynamical evolution of the system under scrutiny. In the case of N gravitating particles, T^*Q will be nothing but $\mathbb{R}^{6N}$. The justification of this fact is straightforward, if we consider what it takes to determine an initial condition $(\mathbf{q}_0, \dot{\mathbf{q}}_0)$: for each particle, we have to specify three numbers that indicate its position and further three numbers that give the velocity vector "attached" to it. We immediately see in what sense this framework naturally fits a substantivalist understanding of space: two N-particle states that agree on all relational facts about the configuration (not only the relative positions but also the relative orientations of the velocities), but disagree on how such a configuration is embedded in Euclidean space, would count as physically distinct possibilities. Then, the natural relationalist move would be to construct a relational configuration space Q_0 by quotienting out from Q all the degrees of freedom associated with an embedding in Euclidean space, such as rigid translations and rotations. If we call $E(3)$ the set of isometries of Euclidean three-space, then $Q_0 = Q/E(3)$: this construction assures us that distinct points in Q that represent the same relational configuration "collapse" to the same point in Q_0. Note that (i) Q_0 admits a well-defined cotangent bundle T^*Q_0, which is equipped with a well-behaved symplectic structure, and (ii) that the starting Hamiltonian defined on Q admits a smooth projection H_0 to T^*Q_0 because it is invariant under the action of $E(3)$.

Belot's theory qualifies as relational, since the ontological facts making up a set of initial data do not encode any notion of position in absolute space or absolute velocity, and the laws of motion specify how these initial data evolve; furthermore, the dynamical laws of the theory are fully defined on relational phase space. However, there are at least three concerns that one can raise about Belot's proposal. In the first place, this theory still has a notion of absolute time inherent in the dynamical laws. In fact, there is nothing in the quotienting out procedure that leads from a dynamics over Q to one over Q_0 that eliminates the absolute temporal metric of Newtonian mechanics, which means that the very same succession of purely relational configurations can unfold at different rates depending on the ticking of an universal external clock. Secondly, there is a clear sense in which spatial relations are Euclidean from the beginning: they are just equivalence classes of embedding degrees of freedom as encoded in $E(3)$. Thirdly, Belot's theory, despite being very close to Newtonian mechanics, is not as empirically predictive as its absolute counterpart. This is obvious because, if we think about all the initial data that are needed in the Newtonian theory, we realize that they must include the rate of change in the orientation of the configuration of N particles with respect to absolute space: in the passage from Q to Q_0, this information is simply washed away. In particular, the Newtonian theory admits models with non-vanishing total angular momentum J of the universe. We repeat that this is not just a metaphysical aspect, but a physical one in the sense that the condition $J \neq 0$ carries with it empirically testable consequences. Belot's proposal to overcome the problem is just to bite the bullet: his relational reduction recovers only part of the Newtonian one, but—given that up to now we have reliable experimental evidence that the total angular momentum J of the universe is zero—it recovers exactly the empirically adequate part.

Let us now turn to Barbour's proposal (see Barbour and Bertotti (1982), Barbour (2003, 2012) for the original resources; Pooley and Brown (2002), sections 6–7, give an excellent overview of the framework, together with some cogent philosophical considerations). Barbour's relationalist motivations are the same as Belot's—that is, to eliminate all the spatial degrees of freedom that produce no observable difference. However, Barbour extends this requirement to temporal degrees of freedom as well. By way of consequence, the construction of his framework involves two steps—namely, the implementation of (i) spatial and (ii) temporal relationalism. As regards the first step, Barbour adopts the same strategy as Belot: he takes standard configuration space Q and quotients out all Euclidean isometries, comprised of scale transformations, which means that he quotients out also the degrees of freedom related to "stretchings" or "shrinkings" of configurations that preserve the ratio of distances. This means that he considers a wider group than $E(3)$—namely, the similarity group $Sim(3)$. Hence, his relational configuration space $Q_0 = Q/Sim(3)$ is aptly called *shape* space, because each configuration in there is individuated by its form and not by its size.

The second step is technically more complicated: firstly, Barbour defines an "intrinsic" difference that measures how much two shapes are similar. This difference is expressed in terms of "best-matching" coordinates. Intuitively, we imagine the two shapes laid down over two distinct Cartesian coordinate grids O and O'; then we hold fixed the first shape and grid and "move" the second by applying transformations in $Sim(3)$ until the two shapes are juxtaposed as close as possible. The best-matching coordinates are then defined as the overlap deficit $O - O'$ between the two coordinate grids. Secondly, he uses this intrinsic metric to define a Jacobi action, thus setting a Jacobi variational principle on Q_0 (see Lanczos (1970), pp. 132–140, for a technical introduction to the Jacobi principle in classical mechanics). The Jacobi action is reparametrization invariant—that is, it does not change whatever "time" parameter we choose.

With this machinery in place, carrying out the variation of the action with respect to the best-matching coordinates, we obtain a set of generalized Euler-Lagrange equations whose integral curves are nothing but the geodesics of Q_0. Given some initial conditions, one of these curves is singled out, which represents the dynamical evolution of the system. This evolution is given in fully relationalist terms: the curve singled out by the equations plus the initial conditions represents a list of relational configurations, which is parametrized by an arbitrary monotonically increasing parameter: hence, there is no external clock that measures dynamical change; on the contrary, it is the change in the list of configurations that enables an (arbitrary) parametrization. The important point is that there exists a particular parametrization of the curve for which the generalized Euler-Lagrange equations take the usual Newtonian form. Thus, if we adopt this (again, arbitrary) parametrization, we obtain a dynamical description that matches the Newtonian one. In this sense, Newtonian mechanics comes out of Barbour's framework by means of something closely resembling a gauge fixing. The descriptive simplicity of the Newtonian formulation then explains why, historically, classical physics was framed in these terms.

There are at least three critical points about this framework worth being highlighted. Firstly, no usual Newtonian potential is compatible with the condition of scale-invariance. Even if it is always possible to reproduce the form of the most usual classical potentials in the appropriate gauge by a clever mathematical manoeuvre, still this mimicking strategy might lead to unwanted physical restrictions, such as no angular momentum exchange between subsystems (see Anderson (2013), section 5.1.2, for a technical discussion of this point). Secondly, the implementation of a geodesic principle on a general shape space might not always be that straightforward. In general cases, in fact, the quotienting out procedure sketched earlier leads to a shape space whose global geometry is that of a stratified manifold, where each "stratum" is a sub-manifold that can differ from the others in many respects, including the dimensionality. It is then quite intuitive to understand that, if $\mathcal{Q}_0$ is a stratified manifold, it is problematic to account for a dynamical evolution given in terms of a geodesic trajectory that hits different strata of $\mathcal{Q}_0$ (see Anderson (2015), section 9.4 and references therein, for discussion). The moral is that Barbour's framework works well in a suitably small region of $\mathcal{Q}_0$, but might break down on a larger scale, depending on the particular geometrical structure of $\mathcal{Q}_0$. Thirdly, as a result of quotienting out the group of rotations from $\mathcal{Q}$, one gets the condition $J = 0$ as a constraint on shape space dynamics. However, even if the actual universe satisfies the condition of $J = 0$, a relationalist theory should be able to account for observable consequences ascribable in absolute terms to a non-vanishing total angular momentum of the universal configuration of matter.

Finally, let us consider the proposal spelled out in Saunders (2013). Like Belot and Barbour, Saunders's aim is to dispense with absolute quantities of motion. However, unlike the former two, he also seeks to save the core conceptual structure of Newton's *Principia*. In order to do so, he shows that the Newtonian laws can be cast in terms of directed distances representing inter-particle separations. This is possible because the absolute notion of "straight trajectory" needed to make sense of inertial motion—which, in turn, is required for rendering Newton's first and second law meaningful— involves too much structure—namely, a privileged affine connection (that of neo-Newtonian space-time). Instead, Newton's laws can make perfect sense even if we replace the talk of straight trajectories with that of relative velocities not changing over time, and this can be accounted for not just by a single preferred connection, but by a whole class of affine connections whose time-like geodesics are mutually non-rotating. This is all that Saunders needs in order to account for accelerations and rotations in relationalist terms: a space-time manifold equipped with enough structure to allow for the comparison of spatial directions (and related angles) at different times (what he calls "Newton-Huygens" space-time; see also Earman (1989), pp. 31–32, for a formal characterization of this space-time). In the Newton-Huygens space-time, unlike the neo-Newtonian one, it is meaningless to talk about the absolute acceleration of a particle, or even its inertial motion (two notions that are tied to a privileged affine connection), while it is perfectly meaningful to ask questions about the change of orientation of a configuration in time.

The huge virtue of this framework is that it is able to recover the full spectrum of Newtonian models, thus accounting also for $J \neq 0$ cases, without invoking absolute notions. Given, in fact, that differences in direction can be defined relationally by admitting a primitive notion of parallelism, and then defining change in direction by comparison of spatial relations at different times, Saunders's theory can account for global rotations in terms of relational quantities. However, the fact that this theory makes it meaningful to compare spatial directions at different times represents a substantial weakening of the relationalist programme. This framework is much less relationalist than Belot's and Barbour's, for which the excision of any physical meaning attached to global rotations is a constitutive feature. In this sense, Saunders's theory is a "halfway house" form of a relationalism, as he himself notes (Saunders (2013), p. 44).

In sum, there are well-grounded reservations whether these relationalist theories fully implement a relationalist ontology. That ontology is relationalist both with respect to space and time, whereas Belot's and Saunders's theories are relationalist only with respect to space. Furthermore, that ontology is not tied to a particular geometry such as Euclidean geometry: a configuration of matter points satisfying axiom 1 and definition 1 of Chapter 2, section 1, does not carry with it any primitive geometrical fact that singles out a distinguished space in which it has to be embedded. By contrast, there is a clear sense in which Belot's and Barbour's spatial relations are Euclidean from the outset: each point in Q_0 can be seen as an equivalence class of Euclidean configurations; there is no way, by fixing a certain gauge, to end up with a configuration embedded in a non-Euclidean space. Also Saunders's relations are inherently Euclidean, since Newton-Huygens space-time encodes the structure of a series of instantaneous 3-dimensional affine spaces equipped with an Euclidean metric. Moreover, all these theories rely on more primitive structure than just distances. The very concept of shape requires primitive facts about angles to be meaningful, so that Barbour's ontology has to include a conformal structure. Saunders's ontology requires not only distances, but *directed* distances, which means that some primitive geometrical facts have to be postulated (especially those making up a standard of space-like parallel transport), which are encoded in an affine structure.

At this point, one may legitimately ask whether it is possible to resort to the Super-Humean strategy to argue that the additional structure of these theories is just part of the package we get when seeking for the best description of motion. The answer is that such a move, in this case, raises substantial worries. Put simply, a Humean justification of the surplus structure would imply that what these theories do is basically to "embellish" a set of relational initial data Δ_0 by embedding them in a more structured set Q_0 and then using the equations of motion cast in terms of this surplus structure to evolve these data until reaching the result Q_t, from which the relational solution Δ_λ would be read off. But that would have unwanted implications. The first of these is that, in this way, neither of the earlier theories could be considered even mildly relationalist anymore, being parasitic on a dynamical description that involves an irreducible surplus structure. In such a case, it would be awkward to prefer these theories to Newtonian gravitation, given that Huggett's Humean strategy works perfectly for the purposes of a relationalist ontology in that context. The second implication is that the strategy of evolving relational data by "stealing a ride" to a non-relationalist dynamics and then discarding the surplus structure as a mere representational means would suspiciously look like a trivial instrumentalist move, as discussed notably by Earman (1989, p. 128) and Belot (2000, p. 10).

In sum, combining a minimalist relationalist ontology with the alternative theory strategy faces a dilemma: either one insists that angles or directions are just part of the Humean package, thus ending up with—to paraphrase Earman (1989, p. 128)—a cheap instrumentalist rip-off of a theory that in any case does not qualify as a genuine relationalist competitor to Newtonian mechanics; or one bites the bullet and introduces more primitive structure in the ontology (be it a conformal or an affine one), thus betraying the original motivations for relationalism from ontological parsimony. That is why we prefer the Super-Humean strategy, as implemented for Newtonian mechanics by Huggett (2006) and Hall (2009, § 5.2).

In our discussion of Newtonian mechanics, we have assumed that matter is made out of a collection of identical matter points that stand in spatial relations with respect to each other. However, classical mechanics has also brought forward very successful theories that describe matter by means of continuous phase space densities evolving in time; consider, for instance, the Vlasov, Navier-Stokes or Euler equations. This rightfully raises the question whether a description of matter in terms of matter points is adequate at all. This question has been answered in the affirmative by Boltzmann who linked Newtonian mechanics and thermodynamics by means of statistical mechanics—that is, a statistical analysis of Newtonian dynamics. Gases, fluids and solids are fundamentally described by a large collection of discrete particles that move according to Newtonian mechanics. However, in the mathematical idealization of the so-called thermodynamic limit,—that is, sending the volume as well as the total particle number to infinity while keeping the density constant—it is much more easy to disregard the particular microscopic and often irrelevant motion of the respective particles and switch to a more coarse-grained description of the motion in terms of phase space densities, which may then be regarded as a description of a cloud of particles. This description, however, does not assign a special ontological status to the phase space density as the latter is just a mathematical quantity derived from the particle positions in order to achieve a convenient and mathematically tractable description of the motion of the particles.

3.2 Bohmian quantum mechanics

When it comes to quantum mechanics, a discussion of ontology makes sense only if one spells out to which version of quantum mechanics one is committed—that is, how one solves the measurement problem. As we mentioned in the introduction, quantum theories that are committed to a distribution of matter in physical space that accounts for measurement outcomes are known as primitive ontology theories (see Allori et al. (2008)). The quantum theory going back to de Broglie (1928) and Bohm (1952a) is the most prominent of them. It is based on a primitive ontology of permanent point particles as in classical mechanics. In the following, we draw on the predominant contemporary formulation of Bohm's theory, known as Bohmian mechanics (see Dürr et al. (2013b)). This theory conceives a dynamical law for the motion of the particles in physical space (and not just for the quantum state, as does the Schrödinger equation). It can be cast as a quantum theory of classical gravitation (since, however, the relevance of gravitation in the realm of light elementary particles might seem questionable—and we use it only for pointing out the parallels with its classical version—one should rather think of this theory as describing a gas of electric charges by replacing the gravitational constant G with Coulomb's constant $(4\pi\epsilon_0)^{-1}$ and the gravitational masses m_k with the electric charges e_k).

In Bohmian mechanics, the t-dependence of v_t is given as functional of a map $\Psi_{(\cdot)} : \mathbb{R} \times \mathbb{R}^{3N} \to \mathbb{C}, (t, q) \mapsto \Psi_t(q)$ by

$$v_{k,t}(Q_t) := \frac{\hbar}{M_k} \, \text{Im} \, \frac{\nabla_k \Psi_t(Q_t)}{\Psi_t(Q_t)}, \tag{3.6}$$

where Ψ_t is requested to be a square-integrable solution to the Schrödinger equation

$$i\hbar \partial_t \Psi_t = H\Psi_t, \tag{3.7}$$

for the operator

$$H = \sum_{k=1}^{N} -\frac{\hbar^2}{2M_k} \Delta_k + V. \tag{3.8}$$

Here, $\hbar \in \mathbb{R}^+$ denotes an additional constant while G, M_k and V are the same mathematical entities as in the preceding section. For a sufficiently regular Ψ_0, equation (3.7) admits a unique solution

$$\Psi_t := e^{-itH}\Psi_0 \tag{3.9}$$

fulfilling $\Psi_t|_{t=0} = \Psi_0$ and, therefore, together with (3.6), yields a unique velocity field

$$v_t(Q_t) = (v_{1,t}(Q_t), ..., v_{N,t}(Q_t)) \qquad \text{for}$$
$$v_{k,t}(Q_t) := \frac{\hbar}{M_k} \, \text{Im} \, \frac{\nabla_k (e^{-itH}\Psi_0)(Q_t)}{(e^{-itH}\Psi_0)(Q_t)}. \tag{3.10}$$

These additional parameters—that is, the relationships described in (3.6), (3.7), and (3.4), Planck's constant $\hbar$, Newton's gravitational constant G, the inertial masses M_k, and the initial wave function Ψ_0—make up the dynamical structure. They are the only degrees of freedom left in the choice of a velocity field v_t admissible in this Bohmian setup. Again, thanks to (3.2), specifying these parameters together with the initial configuration $Q_0 \in \mathbb{R}^{3N}$ determines the motion $t \mapsto Q_t$ uniquely.

Note the similarity between classical mechanics and Bohmian quantum mechanics as set out here. Both theories fit into the framework of the first order differential equation (3.2), which requires the specification of the velocity field v_t. All admissible candidates for v_t can be found with the help of only a few additional parameters and the relations—cf. (3.3) and (3.6)–(3.7), respectively—which rule the t-dependence of v_t – cf. (3.5) and (3.10), respectively. Specifying further parameters—here the initial wave function Ψ_0 instead of the initial velocities $\dot{q}_{k,0}$—uniquely determines the velocity field v_t, which in turn, for a given initial configuration Q_0, uniquely determines the motion $t \mapsto Q_t$ as solution of (3.2).

The fact that in classical mechanics v_t itself is usually defined by specifying its derivative as in (3.3), which effectively results in a second order equation, concerns only the way of formulating the dynamical structure but does not affect the ontology. The same is possible for Bohmian mechanics: Bohm (1952a) conceived the velocity law for the particles as a second order equation. Doing so does not change the role of the introduced quantities: given v_t, the fundamental law is (3.2). This law describes the change in the configuration of matter points Q_t. The ontology that we propose consists in the configuration of matter points individuated by distance relations and the change of these relations. The dynamical parameters that a physical theory introduces are mathematical variables that are employed in order to find a good candidate for v_t. Thus, beside the other constants that have to be fixed, the initial velocities in classical mechanics are just parameters that help to easily find an admissible v_t thanks to (3.5). The role of the initial velocities is taken over by the initial wave function in quantum theory. In the same vein, besides the other constants, it determines the admissible v_t by (3.10). Therefore, also the initial wave function just is a mathematical variable that helps to carry out the task of finding a suitable candidate for v_t. In the second order formulation of Bohmian mechanics, the initial velocities $\dot{q}_{k,0}$ appear as additional parameters. But these parameters are redundant, since, in order to reproduce the statistical predictions of quantum mechanics, they have to be fixed in accordance with (3.6); there is no empirical data that would justify or require a different choice.

On this view, hence, forces are not explanatory in the sense that they describe agent-like entities that push the particles around. Consequently, from the perspective of a parsimonious ontology of matter points and change in their configuration, it makes no sense to argue that the second order formulation of Bohmian mechanics in terms of forces, including a specific quantum force, is explanatory superior to the first order formulation of Dürr et al. (2013b) on which we rely here (see Belousek (2003) for that debate). In a nutshell, the formulation of Bohmian mechanics in terms of forces does not add explanatory value, but, to the contrary, creates new drawbacks, since the forces would have to coordinate the motion of the particles instantaneously all over space instead of propagating in space (cf. Chapter 2, section 3, and in particular the combination of Bohmian mechanics and Humeanism about dynamical structure set out there; by contrast, see Suárez (2015) for a proposal for an ontology for Bohmian mechanics in terms of a multitude of dispositions instantiated by each individual particle).

Nonetheless, the wave function Ψ_t may strike one as an odd dynamical parameter in an ontology of matter points that are characterized only by their distance relations, so that any dynamical parameter is there to capture the change in the distance relations: the wave function does not care about the actual particle positions—that is, the distance relations among the matter points. Moreover, in quantum physics, all the dynamical parameters are situated on the level of the wave function, including mass and charge. That notwithstanding, as is evident from equation (3.10), the wave function has the job to yield a velocity field along which the particles move as output, given any possible configuration of matter points as represented by a point in configuration space as input. In other words, in being defined for any possible particle configuration admitted in configuration space, the wave function enables us to obtain the velocity of the actual particle configuration as output, if that configuration is given as input. Consequently, although all the dynamical parameters are situated on the level of the wave function, it would be wrong-headed to say that the interaction takes place on the level of the wave function: interaction is the correlated change of the particles' velocities. That correlated change in physical space is represented by means of the wave function in configuration space and the dynamical parameters that are situated on its level. The wave function then serves to determine the velocity of the particles, as described in equation (3.10), whereby the interactions between the matter points enter into the theory via the Hamiltonian operator (3.8). Since the equation of motion for the particles is a first order differential equation, it has a unique solution for any initial particle configuration (assuming the vector field generated by the wave function is sufficiently "nice"). In this sense, one can say that Bohmian mechanics replaces the specification of an initial velocity in classical mechanics with the specification of an initial wave function. Given the wave function and the particle configuration at a certain time, their evolution is fixed for all times.

Let us now briefly consider how the Super-Humean and the alternative theory strategy fare in Bohmian mechanics. As regards the former, we already mentioned the Humean treatment of the wave function in Chapter 2, section 3. Space-time in Bohmian mechanics poses no problem for Humeanism, given that the laws of Bohmian dynamics, although being different from the Newtonian ones, are nonetheless formulated over an absolute Newtonian spatio-temporal background: the manner in which Huggett (2006) deals with space-time in the Newtonian case can simply be applied to the Bohmian case. As regards the alternative theory strategy, the proposal of Belot (1999) would require forcing the Bohmian formalism in a Hamiltonian context; this is possible, but results in a quite unreasonably complicated formalism with a huge amount of surplus descriptive structure, which is not needed by Bohmian dynamics (see Holland (2001a,b) for a decently worked out Hamiltonian version of Bohmian mechanics). Barbour's framework, by contrast, can in a natural way be applied to Bohmian mechanics (see Vassallo (2015), Vassallo and Ip (2016)). The same goes for the milder relationalist approach of Saunders (2013) since this approach would basically amount to re-write the Bohmian theory in a way that makes it meaningless to refer distances and directions to any point taken as "origin".

Bohmian mechanics illustrates very clearly the point made at the beginning of Chapter 2, section 1—namely, that even the classical parameters mass and charge are situated on the level of the wave function. Consider the following experiment: imagine a charged particle whose (effective) wave function is of the form $\psi = \phi_A + \phi_B$, where ϕ_A and ϕ_B are of equal size and shape but concentrated on two distant regions of space that we denote by A and B, respectively. (These regions could be surrounded by infinite high potential walls— or, more simply put, a box—to keep the wave function from spreading.) The particle is located in one of these regions, let's say in A. Not surprisingly, the trajectory of a second charged particle passing near A is affected by the electromagnetic interactions and deflected towards A, if it has opposite charge, or away from A if it has equal charge as particle one. However, if that second particle were passing near region B, it would be affected *in the very same way*, no matter how far away that is from the actual position of the other particle. This scenario demonstrates, firstly, the explicitly non-local character of Bohmian mechanics. It also shows that it would be wrong to think of charge in the familiar way as something localized at the position of the particles. A similar reasoning would apply to the particle mass, in so far as gravitational interactions play a role in quantum mechanics (see Brown et al. 1995, 1996; Pylkkänen et al. 2015).

Furthermore, Bohmian mechanics illustrates that it is a contingent matter of fact that the dynamical structure represents the matter points as being sorted into different particle species. One can take the contingency of this fact into account by formulating the physical theory in such a way that not only the primitive ontology but also the dynamical structure is not committed to different particle species from the outset. Goldstein et al. (2005a,b) have shown how to do this for Bohmian mechanics. They reformulate Bohmian mechanics in such a way that its dynamics reflects the ontological commitment to propertyless particles, treating all particles as *identical*. To appreciate what this means and how the reformulation is carried out, let us begin with the following observation. If we insist that particles are distinguished only by their positions instead of by intrinsic properties, we note that the configuration space $\mathbb{R}^{3N}$ has too much mathematical structure in that it cares about permutations of the particle labels. That is to say the following: the nature of the Bohmian law of motion (being a first-order differential equation on configuration space) is such that it determines at every time t the change of the system's spatial configuration depending on the current configuration $Q(t)$. However, there are neither intrinsic properties nor relations distinguishing the configuration represented by the tuple $(Q_1, Q_2, \ldots, Q_N)$ from, say, the configuration represented by the tuple $(Q_2, Q_1, \ldots, Q_N)$ with the particles 1 and 2 interchanged. It is thus understood that—for so-called *identical* or *indistinguishable* particles—the natural configuration space of an N-particle system is not $\mathbb{R}^{3N}$, but

$$^N\mathbb{R}^3 := \{S \subseteq \mathbb{R}^3 \mid \sharp S = N\}, \qquad (3.11)$$

which is the set of all subsets of $\mathbb{R}^3$ containing exactly N elements.

Consequently, the wave function of the system should now be defined on the configuration space $^N\mathbb{R}^3$ as well, which in fact can be done (see Goldstein et al. (2005a), section 4). Nevertheless, it is still more convenient, in general, to represent the quantum state as a function on $\mathbb{R}^{3(N)}$ (which can be regarded, mathematically, as the universal covering space of $^N\mathbb{R}^3$). As long as we consider a system in which all particles are associated with the same mass and charge, the demand of consistency then leads immediately to a wave function that is symmetric or anti-symmetric under permutations of the particle coordinates and hence to the famous boson/fermion alternative. In Dürr et al. (2006), $^N\mathbb{R}^3$ was thus already introduced as the configuration space of identical or indistinguishable particles, referring to a *single* species of particles, and it is shown how the quantum statistics of identical particles thus arise in the Bohmian theory (see also Dürr and Teufel (2009), ch. 8.5).

However, we now note that as soon as we have to admit more than one value for the parameters M_k, the standard formulation of Bohmian mechanics breaks down. That is because equation (3.6) no longer defines a law of motion on $^N\mathbb{R}^3$, since it discriminates different particles by their associated mass, while configurations represented on $^N\mathbb{R}^3$ do not do so. The basic idea of Goldstein et al. (2005a,b) is thus to symmetrize equation (3.6) in order to get a permutation invariant equation, because any permutation invariant equation on $\mathbb{R}^{3N}$ defines, in a canonical way, a law of motion on $^N\mathbb{R}^3$, the configuration space of *identical* particles. In this way, they show that we can treat *all* particles as identical, while still accounting for the empirical data that, as usual, are explained in terms of a particle "zoo".

To preserve equivariance of the law—i.e. the conservation of total probability by the Bohmian flow—the symmetrization has to be done in the following way. For the rest of this section, we adopt the notation of Goldstein et al. (2005a,b) to avoid mixing the aforementioned label-dependent definitions with the label-independent approach here. The standard guiding equation (3.6) can be written in the form

$$\frac{dQ}{dt} = \frac{j(Q(t))}{\rho(Q(t))},$$
(3.12)

where

$$\rho = \psi^*\psi$$

is the probability density and $j = (j_1, \ldots, j_N)$ with

$$j_i = \frac{\hbar}{M_i} \operatorname{Im} \psi^* \nabla_i \psi$$

the probability current corresponding to the system's wave function ψ. In equation (3.12), numerator and denominator have to be symmetrized independently by summing over all possible permutations of the particle labels $1, \ldots, N$. Hence, we get a new, permutation invariant guiding equation, which reads

$$\frac{dQ_k}{dt} = \frac{\sum_{\sigma \in S_N} j_{\sigma(k)} \circ \sigma}{\sum_{\sigma \in S_N} \rho \circ \sigma} (Q(t)).$$
(3.13)

Here, the sum goes over all elements of the permutation group S_N and

$$\sigma Q := \left(Q_{\sigma(1)}, \ldots, Q_{\sigma(N)} \right)$$

means that every coordinate Q_i is assigned a new index $Q_{\sigma(i)}$, changing the order in the N-tupel.

In this theory, which Goldstein et al. (2005a,b) dubbed *identity-based Bohmian mechanics*, we do not attribute *a priori* any mass to any specific particle. The law of motion merely determines N trajectories for N particles, and it is a characteristic of this law that one of those trajectories happens to behave—at least in the relevant circumstances—like the trajectory of a particle with mass M_1, another like the trajectory of a particle with mass M_2, and so on, depending only on the (contingent) initial conditions of the system, respectively the universe.

To illustrate how this works, let us discuss an example given in Goldstein et al. (2005a, section 3), which compares the standard formulation of Bohmian mechanics with the identity-based version. Consider a two-particle universe consisting of an electron with mass M_e and a muon with mass M_μ. Suppose, for simplicity, that they are in a non-entangled state $\Psi(q_1, q_2) = \phi(q_1)\chi(q_2)$ (note that we could symmetrize this wave function, though this would be redundant when plugged into the symmetrized guiding equation). Then, the standard guiding law (3.6) leads to the following equations of motion:

$$\frac{dQ_1}{dt} = \frac{\hbar}{M_e} \operatorname{Im} \frac{\nabla\phi(Q_1)}{\phi(Q_1)},$$

$$\frac{dQ_2}{dt} = \frac{\hbar}{M_\mu} \operatorname{Im} \frac{\nabla\chi(Q_2)}{\chi(Q_2)}.$$

$$(3.14)$$

In contrast, the symmetrized guiding equation (3.13) reads

$$\frac{dQ_1}{dt} = \frac{\frac{\hbar}{M_e}|\chi(Q_2)|^2 \operatorname{Im}(\phi^*(Q_1)\nabla\phi(Q_1)) + \frac{\hbar}{M_\mu}|\phi(Q_2)|^2 \operatorname{Im}(\chi^*(Q_1)\nabla\chi(Q_1))}{|\phi(Q_1)|^2|\chi(Q_2)|^2 + |\phi(Q_2)|^2|\chi(Q_1)|^2},$$

$$\frac{dQ_2}{dt} = \frac{\frac{\hbar}{M_\mu}|\phi(Q_1)|^2 \operatorname{Im}(\chi^*(Q_2)\nabla\chi(Q_2)) + \frac{\hbar}{M_e}|\chi(Q_1)|^2 \operatorname{Im}(\phi^*(Q_2)\nabla\phi(Q_2))}{|\phi(Q_1)|^2|\chi(Q_2)|^2 + |\phi(Q_2)|^2\chi(Q_1)|^2}.$$

$$(3.15)$$

We see that equation (3.14) can easily be taken to suggest that there is an intrinsic mass and thus a distinct type to every particle: particle 1, described by the coordinates Q_1, is the electron with mass M_e, while particle 2, described by the coordinates Q_2, is the muon with mass M_μ. In equation (3.15), by contrast, neither Q_1 nor Q_2 is designated as the position of the electron, respectively the muon. *A priori*, the two particles are distinguished only by the position that they have at time t. However, if we consider a situation in which ϕ and χ have disjoint support, say, when one wave packet is propagating to the left and the other one to the right, one of the two sums in the nominators and denominators will be zero, so that the equation of motion effectively reduces to equation (3.14) (possibly with the indices 1 and 2 interchanged). This is to say, in particular, that in situations where the two-particle wave function is suitably decohered, one of the particles will play the role of the electron—being effectively described by equations (3.6) and (3.7) with the parameter M_e, while the other one will play the role of the muon—being effectively described by equations (3.6) and (3.7) with the parameter M_μ.

Which trajectory turns out to be guided by which part of the wave function thereby depends only on the law of motion and the (contingent) initial conditions of the system, rather than on intrinsic properties of the particles. In fact, if both parts of the wave function were brought back together and then separated again, one and the same particle could switch its role from being the electron to being the muon, and *vice versa*. Hence, to be an electron, a muon, or a positron, etc., is nothing more than to move—in the relevant circumstances—electronwise, muonwise, or positronwise, and so forth.

Apart from such circumstances in which the different parts of the wave function are well separated, one could say that the particles in the previous example are guided by a superposition of (what one would usually call) an electron wave function and a muon wave function. However, it would be misleading to claim that this amounts to a superposition of being an electron and being a muon. Ontologically, there are no superpositions of anything, only propertyless particles moving on definite trajectories. Rather, the labels "electron", "muon", etc., are meaningless in the general case.

One obvious objection to the move proposed by Goldstein et al. (2005a, b) is that the guiding law (3.13) is much more contrived than the one in standard Bohmian mechanics. This is again an illustration of the fact that simplicity in ontology and simplicity in representation pull in opposite directions. That notwithstanding, a few things can be said to address the concern of buying into too complicated a formalism. First, one should note that the apparent complexity of equation (3.13) is really just the price for expressing a law of motion for configurations in $^N\mathbb{R}^3$ on the coordinate space $\mathbb{R}^{3N}$ and does not automatically amount to more complicated physics. Second, it should be noted that (modulo some subtleties discussed by Goldstein et al. (2005a,b)), the symmetrized theory will give rise to the familiar statistical description of subsystems in terms of effective wave functions, which is all that matters for most practical purposes.

In this context, it should also be noted that, given the universal wave function, the "right" statistical description of subsystems—that is, the one agreeing with the predictions of standard quantum mechanics, arises for typical initial conditions in terms of the particle configuration—that is, in quantum equilibrium (see section 4). Hence, the emergence of different particle types as empirically observed in nature is not attributed to special initial conditions (quite the opposite), though it is explained by the particular form of the universal wave function.

Finally, concerning the (empirical) content of the proposed theory, it should be emphasized that the trajectories described by identity-based Bohmian mechanics will in general differ from those obtained from standard Bohmian mechanics, but that the statistical predictions for experimental outcomes are the same. In this sense, the symmetrized theory is empirically equivalent to Bohmian mechanics and hence empirically equivalent to standard quantum mechanics.

3.3   The GRW quantum theory

When it comes to the ontology of quantum physics, the argument for endorsing Bohmian mechanics is not that its ontology matches the one of classical mechanics, but that it provides the best solution to the measurement problem. As mentioned in the introduction, we take the arguments by Bell (2004, ch. 7) and Maudlin (2010, 2015) among others for solving this problem in terms of what is known as a primitive ontology of a spatial configuration of matter and its evolution in physical space to be convincing. This strategy accounts for the experimental evidence in terms of there always being a well-defined configuration of matter in physical space; the quantum state then is the means to represent the evolution of that configuration (see notably Allori et al. (2008)). In other words, there are no superpositions of anything in physical space, although, of course, the dynamics of the evolution of the configuration of matter in physical space is not classical. Bohmian mechanics is the most well known and the most elaborate implementation of this strategy.

Nonetheless, if one pursues this strategy, there are other options available than endorsing a primitive ontology of permanent particles. In particular, instead of subscribing to a discrete ontology of matter points, one can try out a continuous ontology of gunk—that is, a continuous matter density field. Let us therefore consider now what is known as the GRWm theory. This theory combines the quantum dynamics proposed by Ghirardi, Rimini and Weber (GRW) (see Ghirardi et al. (1986)) and further developed by Ghirardi et al. (1990) (see also Gisin (1984, 1989)) with the primitive ontology of a matter density field stretching out throughout an absolute background space (see Ghirardi et al. (1995)).

In GRW, the evolution of the wave function Ψ_t is given by a modified Schrödinger equation. The latter can be defined as follows: the wave function undergoes spontaneous jumps at random times distributed according to the Poisson distribution with rate $N\lambda$. Between two successive jumps the wave function Ψ_t evolves according to the usual Schrödinger equation. At the time of a jump the kth component of the wave function Ψ_t undergoes an instantaneous collapse according to

$$\Psi_t(x_1, ..., x_k, ..., x_N) \mapsto \frac{(L_{x_k}^x)^{1/2}\Psi_t(x_1, ..., x_k, ..., x_N)}{\| (L_{x_k}^x)^{1/2}\Psi_t \|}, \tag{3.16}$$

where the localization operator $L_{x_k}^x$ is given as a multiplication operator of the form

$$L_{x_k}^x := \frac{1}{(2\pi\sigma^2)^{3/2}} e^{-\frac{1}{2\sigma^2}(x_k-x)^2}, \tag{3.17}$$

and x, the centre of the collapse, is a random position distributed according to the probability density $p(x) = \| (L_{x_k}^x)^{1/2}\Psi_t\|^2$. This modified Schrödinger evolution captures in a mathematically precise way what the collapse postulate in textbook quantum mechanics introduces by a *fiat*—namely, the collapse of the wave function so that it can represent localized objects in physical space, including in particular measurement outcomes. GRW thereby introduce two additional parameters, the mean rate λ as well as the width σ of the localization operator, which can be regarded as new constants of nature whose values can be inferred from (or are at least bounded by) experiments (such as chemical reactions on a photo plate, double slit experiments, etc.). An accepted value of the mean rate λ is of the order of $10^{15}s^{-1}$. This value implies that the spontaneous localization process for a single particle occurs only at astronomical time scales of the order of $10^{15}s$, while for a macroscopic system of $N \sim 10^{23}$ particles, the collapse happens so fast that possible superpositions are resolved long before they would be experimentally observable. Moreover, the value of σ can be regarded as localization width; an accepted value is of the order of $10^{-7}m$. The latter is constrained by the overall energy increase of the wave function of the universe that is induced by the localization processes.

However, it is obvious that modifying the Schrödinger equation is, by itself, not sufficient to solve the measurement problem: to do so, one has to answer the question of what the wave function and its evolution represents. One therefore has to add to the GRW equation a link between the evolution of the mathematical object Ψ_t in configuration space and the distribution of matter in physical space in order to account for the outcomes of experiments and, in general, the observable phenomena. Ghirardi et al. (1995) accomplish this task by taking the evolution of the wave function in configuration space to represent the evolution of a matter density field in physical space. This then constitutes what is known as the GRWm theory and amounts to introduce in addition to Ψ_t and its time evolution a field $m_t(x)$ on physical space $\mathbb{R}^3$ as follows:

$$m_t(x) = \sum_{k=1}^{N} M_k \int d^3x_1 ... d^3x_N \, \delta^3(x - x_k)|\Psi_t(x_1, ..., x_N)|^2. \qquad (3.18)$$

This field $m_t(x)$ is to be understood as the density of matter in physical space $\mathbb{R}^3$ at time t (see Allori et al. (2008), section 3.1). Hence, on this theory, despite its formulation in terms of particle numbers, there are no particles in the ontology. More generally speaking, there is no plurality of fundamental physical systems. There is just one object in the universe—namely, a matter density field that stretches out throughout space and that has varying degrees of density at different points of space, with these degrees of density changing in time.

By introducing two new dynamical parameters—lambda and sigma—whose values have to be put in by hand, the GRW theory abandons the simplicity and elegance of the Schrödinger equation and the Bohmian guiding equation, without amounting to a physical benefit (there is of course a benefit in comparison to stipulating the collapse postulate by a simple *fiat*, but doing so is no serious theory). Indeed, there is an ongoing controversy whether the GRWm ontology of a continuous matter density field that develops according to the GRW equation is sufficient to solve the measurement problem. The reason is the so-called problem of the tails of the wave function. This problem arises from the fact that the GRW theory mathematically implements spontaneous localization by multiplying the wave function with a Gaussian, such that the collapsed wave function, although being sharply peaked in a small region of configuration space, does not actually vanish outside that region; it has tails spreading to infinity. In the literature starting with Albert and Loewer (1996) and P. Lewis (1997), it is therefore objected that the GRW theory does not achieve its aim—namely, to describe measurement outcomes in the form of macrophysical objects having a definite position. However, there is nothing indefinite about the positions of objects according to GRWm. It is just that an (extremely small) part of each object's matter is spread out through all of space. But since the overwhelming part of any ordinary object's matter is confined to a reasonably small spatial region, we can perfectly well express this in our (inevitably vague) everyday language by saying that the object is in fact located in that region (see Monton (2004), pp. 418–419, and Tumulka (2011)). Thus, the GRWm ontology offers a straightforward solution to what Wallace (2008, p. 56) calls the problem of bare tails.

However, there is another aspect, which is known as the problem of structured tails (see Wallace (2008), p. 56). Consider a situation in which the pure Schrödinger evolution would lead to a superposition with equal weight of two macroscopically distinct states (such as a live and a dead cat). The GRW dynamics ensures that the two weights do not stay equal, but that one of them (e.g. the one pertaining to the dead cat) approaches unity while the other one becomes extremely small (but not zero). In terms of matter density, we then have a high-density dead cat and a low-density live cat. The problem is that it seems that the low-density cat is just as cat-like (in terms of shape, behaviour, etc.) as the high-density cat, so that in fact there are two cat-shapes in the matter density field, one with a high and another one with a low density. There is an ongoing controversy about this problem: Maudlin (2010, pp. 135–138) takes it to be a knock down objection against the GRW matter density ontology, whereas others put forward reasons that aim at justifying to dismiss the commitment to there being a low density that is as cat-like as the high-density cat in the matter density field (see notably Wallace (2014), Albert (2015), pp. 150–154, and Egg and Esfeld (2015), section 3).

Be that as it may, there arguably is another, more important drawback of the GRW dynamics that concerns the meaning of the spontaneous localization of the wave function in configuration space for the evolution of the matter density field in physical space. To illustrate this issue, consider a simple example—namely, the thought experiment of one particle in a box that Einstein presented at the Solvay conference in 1927 (the following presentation is based on de Broglie's version of the thought experiment in de Broglie (1964), pp. 28–29, and on Norsen (2005)): the box is split in two halves which are sent in opposite directions, say from Brussels to Paris and Tokyo. When the half-box arriving in Tokyo is opened and found to be empty, there is on all accounts of quantum mechanics that acknowledge that measurements have outcomes a fact that the particle is in the half-box in Paris.

On GRWm, the particle is a matter density field that stretches over the whole box and that is split in two halves of equal density when the box is split, these matter densities travelling in opposite directions. Upon inter-action with a measurement device, one of these matter densities (the one in Tokyo in the example given earlier) vanishes, while the matter density in the other half-box (the one in Paris) increases so that the whole matter is con-centrated in one of the half-boxes. One might be tempted to say that some matter travels from Tokyo to Paris; however, since it is impossible to assign any finite velocity to this travel, the use of the term "travel" is inappropri-ate. For lack of a better expression let us say that some matter is delocated from Tokyo to Paris (this term has been proposed by Matthias Egg, see Egg and Esfeld (2014), p. 193); for even if the spontaneous localization of the wave function in configuration space is conceived as a continuous process as in Ghirardi et al. (1990), the time it takes for the matter density to dis-appear in one place and to reappear in another place does not depend on the distance between the two places. This delocation of matter, which is not a travel with any finite velocity, is quite a mysterious process that the GRWm ontology asks us to countenance.

On Bohmian mechanics, by contrast, in this example, there always is one particle moving on a continuous trajectory in one of the two half-boxes, and opening one of them only reveals where the particle was all the time. In other words, Bohmian mechanics provides a local account of the case of a particle in a box. However, when moving from Einstein's thought experiment with one particle in a box (1927) to the EPR experiment (Ein-stein et al. (1935)), even Bohmian mechanics can no longer give a local account, as proven by Bell's theorem (Bell (2004), ch. 2; see also notably chs. 7 and 24). On the GRWm theory, again, the measurement in one wing of the experiment triggers a delocation of the matter density, more precisely a change in its shape in both wings of the experiment, so that, in the version of the experiment by Bohm (1951, pp. 611–622) the shape of the matter density constitutes two spin measurement outcomes. On Bohmian mechanics, fixing the parameter in one wing of the EPR experi-ment influences the trajectory of the particles in both wings via the wave function of the whole system, which consists of the measured particles as well as of the particles that make up the measuring devices.

Hence, in this case, it clearly comes out that according to the Bohmian velocity equation (3.10), the velocity of any particle depends strictly speak-ing on the position of all the other particles. However, each particle always moves with a determinate, finite velocity so that its motion traces out a con-tinuous trajectory, without anything jumping—or being delocated—in physical space. The best conjecture for a velocity field that captures this motion that we can make—namely (3.10), requires acknowledging that the motions of these particles are correlated with each other, but this does not imply a commitment to there being some spooky agent or force in nature that instantaneously coordinates the motions of all the particles in the universe. Quantum physics just teaches us that it is a fact about the universe that when we seek to fill in (3.2) with a simple and general law that accounts for the empirical evidence, we have to write down a law that represents the motions of the particles to be correlated with one another (cf. the remarks and references on Bohmian Super-Humeanism in Chapter 2, section 3).

Again, our proposal for a minimalist ontology makes intelligible why correlated motion is in principle to be expected: if all the change is change in the distance relations among matter points in a given configuration of matter points, one cannot change the distance between two matter points without thereby also affecting the distances among in principle all the other matter points in the configuration. Thus, as mentioned in Chapter 2 at the end of section 2, this ontology accounts for non-local correlations without it making sense to call for something that transmits instantaneous particle interaction across space. Nonetheless, this is a general account of correlated motion, not a specific account of the quantum correlations. Since our minimalist ontology accepts the change in the distance relations as a primitive matter of fact (axiom 2 in Chapter 2, section 1), it takes explanations to end in distance relations and their change, arguing that anything beyond distance relations and their change that one endorses in the ontology creates new drawbacks instead of providing deeper explanations.

However, Einstein (1948) is certainly right in pointing out that a complete suspension of the principles of separability and local action would make it impossible to do physics: a theory that says that the motion of any object is effectively influenced by the position of every other object in the configuration of matter of the universe would be empirically inadequate and rule out any experimental investigation of nature. In order to meet Einstein's requirement, it is not necessary to rely on a dynamics of the collapse of the wave function, as does GRW. Bohmian mechanics fulfills this condition because decoherence will in general destroy the entanglement between large and/or distant systems, allowing to treat them, for all practical purposes, as evolving in an independent manner. Moreover, while accounting for all phenomena of non-relativistic quantum mechanics, Bohmian mechanics is able to recover classical behaviour in the relevant regimes (see Dürr et al. (2013b), ch. 5). Since Bohmian mechanics is a theory about the motion of particles, this classical limit does not involve or require any change in the ontological commitment, but consists in the proposition that typical Bohmian trajectories look approximately Newtonian on macroscopic scales (if the characteristic wave length associated to ψ is small compared to the scale on which the interaction potential varies). Altogether, the Bohmian theory, against the background of an ontology of matter points that are characterized only by their relative positions and a dynamics for the change of these positions, illustrates that there is nothing suspicious about a non-local dynamics.

Apart from the matter density ontology, there is another ontology for the GRW theory available. This ontology goes back to Bell (2004, ch. 22, originally published 1987): whenever there is a spontaneous localization of the wave function in configuration space, that development of the wave function in configuration space represents an event occurring at a point in physical space. These point events are known as flashes; the term "flash", however, is not Bell's, but was coined by Tumulka (2006, p. 826). According to the GRW flash theory (GRWf), the flashes are all there is in physical space. Macroscopic objects are, in the terms of Bell (2004, p. 205), galaxies of such flashes. Consequently, the temporal development of the wave function in configuration space does not represent the distribution of matter in physical space. It represents the objective probabilities for the occurrence of further flashes, given an initial configuration of flashes. Hence, there is no continuous distribution of matter in physical space—namely, no trajectories of particles—and no field such as a matter density field either. There only is a sparse distribution of single events. Although GRWf and GRWm are rival proposals for an ontology of the same formalism (the GRW quantum theory), there also is a difference between them on the level of the formalism: if one endorses the GRWm ontology, it is reasonable to develop the GRW equation into a formalism of a continuous spontaneous localization of the wave function (as done in Ghirardi et al. (1990)); by contrast, if one subscribes to the GRWf ontology, there is no point in doing so.

Whereas Bohmian mechanics vindicates an ontology of discrete objects (matter points, particles) in quantum physics and GRWm vindicates an ontology of gunk, it may seem that GRWf vindicates an ontology of super-substantivalism. In order to answer the question of what the flashes are, one may be inclined to say that they are properties of space: in brief, space flashes at some points. However, this is just a linguistic trick in the sense of Sklar (1974, pp. 166, 222–223): one simply changes the language from using substantives that characterize objects in space to verbs that are predicated of points of space. However, the verb "to flash" does not express a *bona fide* property of space, with topological, affine and metrical properties setting the paradigm for what *bona fide* properties of points of space are. It is a placeholder for a characterization of matter that would have to be filled in.

However, one cannot apply to GRWf the answer to this question that we propose for matter points—namely, that their essence are the distance relations in which they stand, since the flashes are ephemeral instead of permanent. Consequently, there is no such thing as the change in distance relations among flashes as the objects that stand in these relations. To put it differently, the GRW flashes are the Bohmian particles deprived of their trajectories, so that there is no possibility to conceive them as the objects with respect to which one represents changing distance relations, with these relations (and their change) making up the essence of these objects. Instead, there is an absolute physical space with flashes coming into existence in it out of nothing and disappearing into nothing.

In any case, the account that the original GRW theory envisages for measurement interactions does not work on the flash ontology—in other words, this ontology covers only the spontaneous appearance and disappearance of flashes, but offers no account of interactions: on the original GRW proposal, a measurement apparatus is supposed to interact with a quantum object; since the apparatus consists of a great number of quantum objects, the entanglement of the wave function between the apparatus and the measured quantum object will be immediately reduced due to the spontaneous localization of the wave function of the apparatus. However, even if one supposes that a measurement apparatus can be conceived as a galaxy of flashes (but see the reservations of Maudlin (2011), pp. 257–258), there is on GRWf nothing with which the apparatus could interact: there is no particle that enters it, no mass density and in general no field that gets in touch with it either (even if one conceives the wave function as a field, it is a field in configuration space and not a field in physical space where the flashes are). There only is one flash (standing for what is usually supposed to be a quantum object) in its past light cone, but there is nothing left of that flash with which the apparatus could interact.

In sum, thus, as the ontology of permanent discrete objects (substances in the guise of particles or matter points) stands out as the best proposal for ontology already on metaphysical reasons alone as argued in Chapter 2, section 1, so the quantum dynamics that Bohmian mechanics bases on this ontology stands out as the most convincing proposal for a solution to the measurement problem in the framework of primitive ontology theories of quantum physics (see Esfeld (2014a) for a detailed comparison of these theories). Of course, this assessment of the dynamics would change if experimental tests of collapse theories like GRW against theories that exactly produce the predictions of textbook quantum mechanics—such as Bohmian mechanics—were carried out successfully and confirmed the collapse theories where they deviate from the standard predictions (see Curceanu et al. (2016) for such experiments).

3.4   Dynamical structure from the universe to subsystems

The dynamical structure that a physical theory introduces is in any case defined for the universe as a whole. To solve equation (3.2)—and the corresponding equations in classical as well as quantum Bohmian mechanics (equations (3.5) and (3.10))—one has to put in initial data for the

whole configuration of matter points. That is to say, the dynamical structure correlates in principle the motion of any matter point with the one
of any other matter point in the universe. As already mentioned in
Chapter 2 at the end of section 2, it is therefore appropriate to speak of
a dynamical holism.

In classical mechanics, as is evident from equation (3.5), one calculates
the velocity for each matter point relative to each other matter point separately, and that velocity depends on the distance between the two matter
points. Nonetheless, to obtain the correct velocity for any given matter
point, one would have to take into account its relation to all the other
matter points. For instance, as soon as there are dynamical parameters
whose value is conserved throughout the universe—such as the total
energy—there are global correlations in the motions of the matter points.
Thus, the difference between classical and quantum mechanics by no
means concerns a difference between a local dynamics and a dynamical
holism. It is this one: the quantum mechanical wave function is defined
only for the configuration of matter as a whole; it correlates the motion
of any matter point with in principle any other matter point *without that
correlation having to depend on the distance between the matter points.*

One can therefore say that the dynamical holism is more obvious in
quantum mechanics than it is in classical mechanics: in quantum mechanics, there is a single dynamical parameter—the wave function—that correlates the motion of all the matter points; in classical mechanics, by contrast,
one attributes dynamical parameters to the matter points taken individually
and figures out their correlated motion pairwise, depending on the distance
between them. However, this difference does not justify regarding quantum
mechanics as endorsing the configuration space on which the wave function
is defined, by contrast to three-dimensional space, as the space in which the
physical reality is situated (see Albert (2015), pp. 142–143, for such a
view). In both cases, dynamical parameters are introduced in order to determine the change in the distance relations among matter points in such a way
that specifying an initial value of these dynamical parameters together with
an initial configuration of matter points is sufficient to fix the whole evolution of the distance relations among the matter points—independently of
whether this is done for the whole configuration at once, or pairwise for
the matter points. In any case, in brief, if the configuration of the universe
Q_0 and the correct velocity field v_t were known, one could deduce from
equation (3.2) a unique motion $t \mapsto Q_t$ of the entire configuration.

That notwithstanding, note that the ontology defined by the two axioms
in section 2.1 has a much wider scope than what is known as the primitive
ontology approach to quantum physics. Notably, it is not tied to three-
dimensional space; the distance relations defining this ontology are not
wedded to a particular geometry. Thus, from the perspective of this ontology, the main objection to the ontology of Albert (2015, chs. 6 and 7) of
only a wave function in configuration space is not that the wave function

only ontology lacks a primitive ontology in terms of objects that are local-ized in three-dimensional space. From the perspective of parsimony, the main objection is the uneconomical dualism of a space defined at least by topological relations and material entities (such as a wave function field) existing on that space that are defined in terms of some intrinsic features, which hence do not come out of the relations constituting that space.

In any case, laws that describe the evolution of the universe as a whole are useless for calculations. Even if we know the types of parameters that enter into the dynamical structure, both classical mechanics of gravitation and Bohmian quantum mechanics suggest not only one but many possible velocity fields v_t, given that for example the initial velocities or the initial wave function, respectively, are free parameters. We cannot know the initial conditions for the configuration of matter as a whole. At best we know a little about a subsystem and nearly nothing about the rest of the universe. Moreover, even if we had precise knowledge of initial conditions, the complexity of calculation increases exponentially with the number of particles of a given system, as is clearly brought out already by the three body problem in classical mechanics. Furthermore, the evolution of a given configuration of matter points may be extremely sensitive to pertur-bations on its initial conditions, so that a slight error about the initial con-ditions may lead to a great error in predicting the evolution of the system. In brief, on the one hand, we seek for universal physical theories and have such theories at our disposal; on the other hand, these theories are useless as they stand when it comes to solving the equations.

What is the way out of this dilemma? Let us use the notation $Q = (Q^{sys}, Q^{env})$ for which Q^{sys} comprises the distance relations that constitute the subsystem under investigation and Q^{env} those of its environment—that is, all the rest of the universe. Instead of asking, given a particular $Q = (Q^{sys}, Q^{env})$ and a velocity field v_t, what is the precise motion $t \mapsto Q_t^{sys}$ to be conducted, we can ask more humbly, with the knowledge that we have, what kind of motion we can expect to happen in *most* cases—that is, for most of the initial configurations Q^{env} and admissible velocity fields v_t. For instance, when flipping a coin n times, it is generically impos-sible to predict the individual outcomes and thus to predict the exact sequence of heads and tails, although this sequence is determined completely by Q_0 and v_t. Nevertheless, it is possible to derive statements like this one: in most cases, the number of heads will be almost equal to the number of tails provided that the number of throws n is large enough. Although we cannot infer precise *deterministic* statements from our physical theory because of our ignorance, it is still possible to infer *probabilistic* ones.

To derive such statements, we first need a measure that tells us what "most" means. This has obviously to be a measure on the space of all pos-sible configurations as well as of the unknown parameters of the dynamical structure. To obtain this measure, we have to define what "most" shall mean at an arbitrary time t. This definition has to be such that it respects a principle of *stationarity* under transport of the equations of motion to any other time t: the notion of "most" must not change in t. Such a measure then enables us to turn predictions of our physical theory concerning subsystems which, due to our ignorance we could not infer, into random variables whose distributions are determined by this measure. Such a measure can be called *typicality* measure. (The term "random variables", however, is somewhat misleading: there is nothing random about these variables, since the macrostate of a system—defined in terms of such variables—is completely determined by its microstate).

To illustrate this procedure, let us come back to classical mechanics and consider a system of N point particles. Denoting by q_i and p_i the position, respectively the momentum of the i'th particle, let us call $X(t) = (q_1(t), \ldots, q_N(t); p_1(t), \ldots, p_N(t))$ the *microstate* of the system at time t. The space of all possible microstates, here $\Gamma := \mathbb{R}^{3N} \times \mathbb{R}^{3N}$, is called *phase space*. Let us now employ the Hamiltonian formulation of the microscopic laws of motion, which take the form

$$\begin{cases} \dot{q}_i = \dfrac{\partial H}{\partial p_i} \\ \dot{p}_i = -\dfrac{\partial H}{\partial q_i} \end{cases}, \tag{3.19}$$

with

$$H(q,p) = \sum_{i=1}^{N} \frac{p_i^2}{2m_i} + V(q_1, \ldots, q_n). \tag{3.20}$$

More compactly, this can be written as

$$(\dot{q}_i, \dot{p}_i) = v^H(q, p), \tag{3.21}$$

where v^H denotes the vector field on Γ generated by the Hamiltonian H. These equations give rise to a Hamiltonian flow $\Phi_{t,0}$ such that $X(t) = \Phi_{t,0}(X)$ for any initial microstate X. In equation (3.20), m_i denotes the mass of the i'th particle and V the interaction potential, which can be split into

$$V(q_1, \ldots, q_n) = \sum_{i<j} V_{int}(q_i - q_j) + V_{ext}(q_1, \ldots, q_N, t). \tag{3.22}$$

V_{int} then corresponds to a pair-interaction among the particles (e.g. gravitation) and V_{ext} is an external potential, summarizing the influences of the environment. Of course, if the N-particle system is the entire universe, then $V_{ext} = 0$, since there is nothing outside the universe.

If V_{ext} is zero (or at least time-independent), such a Hamiltonian system has several nice properties. For one, it conserves the total energy, meaning that $H = const.$ along any solution of (3.19). Furthermore, by Liouville's theorem, the Hamiltonian flow conserves phase space volume. This is to say that the uniform Lebesgue measure λ is a *stationary* measure on Γ in the sense that for all $t \geq 0$ and any Borel set $A \subseteq \Gamma$,

$$\lambda(\Phi_{t,0}A) = \lambda(A). \tag{3.23}$$

For fixed $E \in \mathbb{R}$, it is usually convenient to consider the reduced phase space $\Gamma_E := \{X \in \Gamma : H(X) = E\}$ to which a system with total energy E is confined by virtue of energy conservation. λ then induces a stationary measure λ_E on the hypersurface Γ_E, which is called the *microcanonical measure*. By convention, we normalize this measure to $\lambda_E(\Gamma_E) = 1$. The Hamiltonian formulation of classical mechanics thus brings out that the Lebesgue measure is distinguished as the simplest stationary measure on phase space.

Making use of this formalism, let us consider the stock example of an ideal gas in a box (with perfectly reflecting walls) that will serve as our toy-model for the universe. The number of particles in such a macroscopic system is of the order of Avogadro's constant—that is $N \sim 10^{23}$. Clearly, determining the actual configuration and / or predicting the trajectories for so many particles is a hopeless task, even if the particles are non-interacting as in our example.

Nevertheless, it is possible to make meaningful predictions about this system. For instance, we can ask the following: what is the rate of particles that have a velocity in the x-direction that is approximately v_0 (where v_0 is some arbitrary, positive number)? We can formalize this in terms of the random variable:

$$F(X) := \frac{1}{N}\sum_{i=1}^{N}\chi_{\{v_{i,x}\in[v_0-\delta,v_0+\delta]\}}(X). \tag{3.24}$$

Here, $\delta > 0$ is a small positive number (giving precise meaning to "approximately v_0") and χ is the indicator function, i.e. $\chi_{\{v_{i,x}\in[v_0-\delta,v_0+\delta]\}}$ equals one if $v_{i,x} = \frac{1}{m}p_{i,x}$ lies in the interval $[v_0-\delta, v_0+\delta]$ and zero if it does not. As mentioned earlier, calling such a variable a "random variable" is somewhat misleading, since the state of the system is always fully determined by its microstate. The point is that a great number of microstates X will in general correspond to (approximately) the same value of such a variable F so that F does not disclose the microstate of the system.

Fixing the mean energy per particle to $\frac{E}{N} = \frac{3}{2}k_B T$ (k_B is the Boltzmann constant and T can later be identified as the temperature of the system), it is a mathematical fact that

$$\lim_{\substack{N\to\infty \\ \frac{E}{N}=\frac{3}{2}k_B T}} \lambda_E(\{X \in \Gamma_E : v_{i,x} \in [a,b]\}) = \int_a^b \frac{\exp\left(-\frac{1}{k_B T}\frac{mv^2}{2}\right)}{\left(\frac{2\pi k_B T}{m}\right)^{3/2}}\, dv. \quad (3.25)$$

From this, one can conclude that for any $\epsilon > 0$,

$$\lambda_E\left(\left\{X : \left|\frac{1}{N}\sum_{i=1}^N \chi_{\{v_{i,x}\in[a,b]\}}(X) - \int_a^b \frac{\exp\left(-\frac{1}{k_B T}\frac{mv^2}{2}\right)}{\left(\frac{2\pi k_B T}{m}\right)^{3/2}}\, dv\right| > \epsilon\right\}\right) \quad (3.26)$$

$$\to 0, \quad N \to \infty.$$

The derivation of this result is a more or less elementary exercise in measure theory. The more profound question, however, is what this result actually means.

The function $\rho(v) \propto \exp\left(-\frac{1}{k_B T}\frac{mv^2}{2}\right)$ is called the *Maxwell distribution*. It is a probability measure, describing a distribution of particle velocities. Note that there is actually nothing random about the velocities of particles in a gas. The velocity (as well as the position) of every single particle is comprised in the microstate X whose evolution is described by a deterministic equation of motion. There are possible X for which the actual distribution of velocities in the gas differs significantly from that described by the Maxwell distribution. For instance, there are microstates X for which all particles move with one and the same velocity. Or microstates X for which a few very fast particles account for almost the entire kinetic energy, while all the others are nearly at rest. But these states are (obviously) very special ones. The crucial and remarkable fact expressed by equation (3.26) is that, for large N, the *overwhelming majority* of possible microstates is such that the distribution of velocities in the gas is (approximately) Maxwellian (cf. Boltzmann (1896), p. 252). The "overwhelming majority of microstates" is thereby defined in terms of the stationary measure λ_E. In this sense, the Maxwell distribution constitutes a prediction of the microscopic particle theory as a statistical regularity manifested for typical (initial) configurations.

Note that the role of the microcanonical measure in this argument is only to give precise meaning to "by far the largest number of all possible states"—that is, to provide a well-defined notion of *typicality*. The Maxwell distribution, in contrast, refers to actual statistical patterns—that is, relative frequencies in typical particle ensembles. Hence, it is important to appreciate the fact that while two measures appear in the mathematical equation (3.26), their status is very different (cf. Goldstein (2012)). To make this point clear, let us add the following observations (see Lazarovici and Reichert (2015) for a detailed argument):

1. Since the box in our example exists only once—even more so if it is supposed to be a model for the universe—probabilistic statements about its (initial) microstate have no empirical meaning. The Maxwellian ρ refers to an actual distribution of velocities that exists in the box. The microcanonical measure does *not* refer to an ensemble of boxes, but pertains to a way of reasoning about the box and the physical laws describing it, allowing us to establish that the observed velocity distribution is typical.

2. Also, the microcanonical measure is not supposed to quantify our knowledge and / or ignorance about the microstate of the gas. While it is correct to say, in some sense, that randomness in a deterministic theory is only due to our ignorance regarding initial conditions, it is important to note the very limited degree to which *knowledge, information, credences* or other subjective notions play a role in the analysis. It is an objective fact that for the great majority of microstates, the distribution of velocities in an ideal gas is (approximately) Maxwellian, and it is this objective fact that we take to be explanatory.

3. With respect to a typicality measure, only sets of very large (≈ 1) or very small (≈ 0) measure are meaningful. Therefore, a probability measure has actually too much mathematical structure and the meaning of "typical" would not change, if we changed our measure in a more or less continuous fashion.

An analogous reasoning can be applied to more mundane examples like the before mentioned coin toss. It is a statistical regularity found in our universe that the relative frequency of heads or tails in a long series of fair coin tosses is approximately 1/2. If we agree that a coin toss is guided by the same laws as all other physical processes in the world, this statistical regularity has to be explained on the basis of the fundamental microscopic theory (here: classical mechanics).

Let us denote by F_i the outcome of the i'th coin toss in a long series of N coin tosses. We say that $F_i = 1$ if the outcome is heads and $F_i = 0$ if the outcome is tails. Since classical mechanics is deterministic, the outcome of every single coin toss is actually determined, through the fundamental laws of motion, by the initial state of the universe. Hence, we have $F_i = F_i(X)$ for $X \in \Gamma$ the initial microstate of the Newtonian universe. The functions F_i are obviously (very) coarse-graining. We do not care about the exact configuration of atoms making up the coin; we do not even care about the exact position or orientation of the coin; we only ask which side is up as the coin lands on the floor. This defines our macroscopic observables.

There are possible initial configurations conceivable that would give rise to a universe that looks pretty much like ours, but in which the relative frequency of heads is very different from 1/2: there are possible initial configurations for which *every* coin ever to be tossed will land on heads, or for which tails will come out two out of three times. But such initial conditions are very special ones. In contrast, *typical* initial conditions of the universe— compatible with there being coins and coin tossers in the first place—are such that the relative frequency of heads or tails in a long series of fair coin tosses is approximately 1/2. Formally, the claim is that for any $\epsilon > 0$,

$$\lambda\left(\left|\frac{1}{N}\sum_{i=1}^{N}F_i(X) - \frac{1}{2}\right| > \epsilon\right) \to 0,\ N \to \infty. \tag{3.27}$$

This is to say that if N is sufficiently large, the set of initial conditions for which the relative frequency of heads deviates significantly from 1/2 is extremely small. Such initial conditions are thus not *impossible*, but *atypical*. (3.27) is a law of large numbers statement. The law of large number is what connects probabilities to relative frequencies in typical ensembles. The distinction between the typicality measure and the probability distribution is here, once again, crucial in order to avoid the usual redundancy of explaining probabilities in terms of probabilities.

On this account, *probabilities are objective*. They apply to patterns in the world instead of subjective beliefs. It is a matter of fact that, as the number of coin tosses N becomes very large, *almost all* sequences of coin toss outcomes manifest the pattern of an approximately equal frequency of heads and tails. This matter of fact is independent of what agents believe about the outcomes (although both are linked: it is of course rational to adapt one's beliefs to the patterns in the world).

However, there are many situations in classical mechanics that are not like the coin toss or the molecules in a gas. For example, when we compute the trajectory of a stone thrown on earth, we can, in general, use a simple deterministic equation without being embarrassed by our ignorance regarding the exact initial microstate of the stone or its environment. There are two conditions satisfied here that allow us to do that:

1. The external forces—that is, the influence of the rest of the universe neglected in the computations is very small compared to the attraction between the stone and the earth. This is usually the case because other gravitating bodies are either very far away or have very small mass compared to the earth.
2. The evolution of the relevant macroscopic variable—here, the centre of mass of the stone—is reasonably robust against variations in the

microscopic initial conditions. In other words, small changes in the microscopic initial conditions have only small effects on the trajectory of the stone. This is why our ignorance about the exact position and momentum of every single particle constituting the stone (or the earth, or the person/apparatus throwing the stone) does not prevent us from making reliable predictions about the motion of its centre of mass.

Nonetheless, even in this case, our prediction for the trajectory of the stone is strictly speaking a typicality result. Atypical events in the environment or fluctuations of the particles constituting the stone could lead to very different outcomes. Hence, to be precise, we would have to cast our result about the trajectory of the stone in a form that looks quite similar to the probabilistic statements (3.26) or (3.27). For instance, denoting by $x(t)$ the computed trajectory (depending on the initial position and momentum of the stone) and by $\tilde{x}(t)$ the actual trajectory of the stone (depending on the initial condition X of the universe), we could write:

$$\lambda\left(\left\{X : \sup_{0\leq t\leq T}|\tilde{x}(t) - x(t)| > \epsilon\right\}\right) \approx 0. \qquad (3.28)$$

Still, the stone throw example points to a striking difference between classical and quantum mechanics. In classical mechanics, we often encounter situations in which correlations between a subsystem and its environment become negligible, allowing for a (more or less) deterministic description of the subsystem. In quantum mechanics, by contrast, the generic situation is much more similar to the coin toss or the molecule in a gas, where predictions of statistical patterns are the best we can hope for. This is so even if we settle for a formulation of quantum mechanics with an ontology of particles and a deterministic law of motion, such as Bohmian mechanics. Let us therefore now discuss how probabilities enter Bohmian mechanics.

For a statistical analysis of Bohmian mechanics, we need (a) a sensible typicality measure defined on configuration space and (b) a procedure to get from the fundamental, universal description in terms of the universal wave function to a well-defined description of Bohmian subsystems. In the following, we will largely rely on the accomplishment of these tasks in Dürr and Teufel (2009) and Dürr et al. (2013b) (see Callender (2007) and Maudlin (2007b) for a philosophical analysis). Given the universal wave function, the appropriate notion of typicality for particle configurations is given in terms of the measure with density $|\Psi|^2$. The crucial feature of this measure is that it is *equivariant*, assuring that typical sets remain typical and atypical sets remain atypical under the Bohmian time evolution. More precisely, if $\Phi_{t,0}^{\Psi}$ is the flow on configuration space induced by the guiding equation (3.6), then

$$\mathbb{P}^{\Psi}(A) := \int_A |\Psi_0|^2 \mathrm{d}^{3N}q = \int_{\Phi_{t,0}^{\Psi}(A)} |\Psi_t|^2 \mathrm{d}^{3N}q \qquad (3.29)$$

holds for any measurable set $A \subseteq \mathbb{R}^{3N}$. Equivariance is thus the natural generalization of stationarity for a non-autonomous (time-dependent) dynamics. The $|\Psi|^2$-measure can be proven to be the unique equivariant measure for the Bohmian particles dynamics that depends only locally on Ψ or its derivatives (see Goldstein and Struyve (2007)). In this sense, it is even more strongly suggested as the correct typicality measure for Bohmian mechanics than the Lebesgue measure is in classical mechanics.

Let us now have a closer look at how Bohmian mechanics treats subsystems of the universe. Suppose that the subsystem consists of $n \ll N$ particles. We then split the configuration space into $\mathbb{R}^{3N} = \mathbb{R}^{3n} \times \mathbb{R}^{3(N-n)}$, so that, writing $q = (x, y)$, the x-coordinates describe the degrees of freedom of the subsystem and the y-coordinates describe the possible configurations of the rest of the universe. Analogously, we split the *actual* particle configuration into $Q = (Q^{sys}, Q^{env}) = (X, Y)$, with $Q^{sys} = X$, the configuration of the subsystem under investigation and $Q^{env} = Y$ the configuration of its environment.

In passing from the fundamental, universal theory to a description of the subsystem, we can just take the universal wave function $\Psi_t(q) = \Psi_t(x, y)$ and plug into the y argument the actual configuration $Y(t)$ of the rest of the universe. The resulting

$$\psi_t^Y(x) := \Psi_t(x, Y(t)) \qquad (3.30)$$

is now a function of the x-coordinates only. It is called the *conditional wave function*. In terms of this conditional wave function, the equation of motion for the subsystem takes the form

$$\dot{X}(t) \propto \mathrm{Im} \left. \frac{\nabla_x \psi_t^Y(x)}{\psi_t^Y(x)} \right|_{x=X(t)} \qquad (3.31)$$

to be compared with (3.6). However, since the conditional wave function depends explicitly on $Y(t)$, its time evolution may be extremely complicated and not follow any Schrödinger-like equation. Fortunately, in many relevant situations, the subsystem will dynamically decouple from its environment. The subsystem has an *effective wave function* φ if and only the universal wave function takes the form

$$\Psi(x, y) = \varphi(x)\chi(y) + \Psi^{\perp}(x, y), \qquad (3.32)$$

where χ and $\Psi^{\perp}$ have disjoint y-support and $Y \in$ supp χ, so that in particular $\Psi^{\perp}(x, Y) = 0$ for almost all x. (Note that this is much weaker than assuming that Ψ has a product structure, which is in general not the case.) This means that we can effectively forget about the empty wave packet $\psi^{\perp}(x, y)$ and describe the subsystem in terms of its own independent wave function φ. If we can furthermore assume that the interaction between subsystem and environment is negligible for some time—that is,

$$V_{ext}(x, y)\varphi(x)\chi(y) \approx 0, \tag{3.33}$$

the effective wave function will satisfy its own, autonomous Schrödinger evolution. Such a φ—normalized to $\int |\varphi(x)|^2 dx = 1$—is the Bohmian counterpart of the usual quantum mechanical wave function. It is these effective wave functions that physicists manipulate in laboratories and for which Born's rule is formulated.

For our statistical analysis, we start by considering the conditional measure

$$\mathbb{P}^{\Psi}(\{Q = (X, Y), X \in \mathrm{d}^n x\}|Y) = \frac{|\Psi((x, Y))|^2 \mathrm{d}^n x}{\int |\Psi((x, Y))|^2 \mathrm{d}^n x} = |\psi^Y(x)|^2 \mathrm{d}^n x. \tag{3.34}$$

In the special situations described by (3.32), the conditional wave function ψ^Y on the right-hand side becomes the effective wave function φ. For practical purposes, though, conditioning on the configuration Y is much too specific, since we have only very limited knowledge of Y. However, many different Y will yield one and the same effective wave function for the subsystem. Collecting all those Y, and using the fact that by yielding the same effective wave function they also yield the same conditional measure (3.34), a simple identity for conditional probabilities yields

$$\mathbb{P}^{\Psi}(\{Q = (X, Y), X \in \mathrm{d}^n x\}|\psi^Y = \varphi) = |\varphi|^2 \mathrm{d}^n x. \tag{3.35}$$

From this formula, one can now derive law of large numbers estimates of the following kind: at a given time t, consider an ensemble of M identically prepared subsystems with effective wave function φ. Denote by X_i the actual configuration of the i'th subsystem. Let $A \subseteq \mathbb{R}^{3n}$ consider the corresponding indicator function $\chi_{\{X_i \in A\}}$, which is 1, if the configuration X_i is in A and 0 otherwise. Then it holds for any $\epsilon > 0$ that

$$\mathbb{P}_t^{\Psi} = \left(\left\{ Q : \left| \frac{1}{N} \sum_{i=1}^{N} \chi_{\{X_i \in A\}}(Q) - \int_A |\varphi(x)|^2 \right| < \epsilon \right\} \right) \to 0, N \to \infty. \tag{3.36}$$

This is to say that for *typical* configurations of the universe, the particles in an ensemble of subsystems with effective wave function φ are distributed according to $|\varphi|^2$. Thus, Born's rule holds in typical Bohmian universes—that is, in *quantum equilibrium*.

Once again, we emphasize that the $|\Psi|^2$-measure given in terms of the *universal* wave function is only used to define typicality. It is *not* supposed to describe an actual distribution of configurations—that is, an ensemble of universes, because the universe exists only once. By contrast, the $|\varphi|^2$-measure on the right-hand side, defined in terms of the effective wave function, does refer to actual particle distributions in a typical ensemble of identically prepared subsystems. Born's rule is thus predicted and explained by Bohmian mechanics.

Comparing equation (3.35) to (3.26) (and recalling the reasoning that led to the respective equations) we recognize the analogy between the derivation of Maxwell's distribution in classical mechanics and Born's rule in Bohmian mechanics. In essence, it is Boltzmann's statistical mechanics applied to two different theories. The status of probabilities and the role of typicality is the same in both cases, although the dynamical laws are strikingly different. On the one hand, this illustrates the deepness and universality of Boltzmann's insights. On the other hand, it shows that there is no need to look for a fundamentally new kind of randomness in the quantum realm. If the microscopic laws and the ontology of the theory are clear, probabilities in quantum mechanics are no more mysterious than they are in classical mechanics.

However, why then does the quantum realm appear to us so much more random and unpredictable than the classical realm? The answer to this question is in part trivial. Quantum mechanics is usually employed to make predictions about microscopic systems, while classical mechanics is most often employed to make predictions about macroscopic systems and coarse-grained observables. The latter is bound to be more robust against our ignorance regarding microscopic initial conditions. Furthermore, our ability to describe a particular subsystem and the level of detail that we can thereby achieve depends heavily on the strength of correlations between the investigated subsystem and the rest of the universe. Newtonian mechanics is a non-local theory, though only in a rather mild sense. Forces fall off quickly with increasing distance (and gravity is very weak to begin with) so that parts of the universe can often be described as autonomous Newtonian systems for all practical purposes.

In quantum mechanics, as clearly brought out by Bohmian mechanics, however, there are correlations possible that do not depend on the distance of the systems. An example is the two particles in the EPR experiment which produce the famous anti-coincidences. This difference stems from the different manners in which these theories define the velocity field. In classical mechanics, as mentioned earlier, the time evolution of the velocity is given separately for each component $k = 1, \ldots, N$ in (3.5). In Bohmian mechanics, the time evolution of the velocities (3.6) is determined by the wave function Ψ_t whose temporal development is given by the Schrödinger evolution (3.7). Both are defined on the whole configuration space. That is why correlations that are independent of the distance of the systems are admitted in quantum physics. Consequently, it is much more difficult to consider any proper part of a Bohmian universe as "isolated", while ignoring the influence of the rest of the universe. As a matter of fact, it is often possible to provide an autonomous Bohmian description of a Bohmian subsystem in terms of an effective wave function. This autonomy, however, can be somehow deceiving, because the effective wave function still depends implicitly on the configuration of the environment (e.g. on the procedure used to prepare that state in an experimental situation).

More precisely (and more profoundly), our possible knowledge about the particle configuration in a Bohmian subsystem is restricted by the theorem of *absolute uncertainty*, which has no analogue in classical physics (see Dürr et al. (2013b), ch. 2). Absolute uncertainty is a direct consequence of the conditional probability formula (3.34): all our *records* about the particle positions—brain states, computer prints, pointer position, etc.—are included in the configuration Y of the rest of the universe. Hence, all possible correlations between these records and the configuration of the subsystem are already taken into account in equations (3.34) and (3.35) that yield Born's rule for the distribution of particle positions.

This connection between our epistemic state and the effective wave function of the subsystem then works in two ways. On the one hand, it means that given a Bohmian subsystem with effective wave function φ, our information about the particle configuration cannot be more precise than what is given by the $|\varphi|^2$-distribution. On the other hand, it means that if we perform additional measurements to determine the particle positions with greater accuracy, the system's effective wave function becomes more and more peaked. Hence, the gradients in the velocity formula (3.6) induce higher and higher possible velocities, depending on the precise initial configuration of the particles. Less uncertainty about the initial particle positions thus implies more uncertainty about the (asymptotic) velocities—this is the source of Heisenberg's uncertainty principle. Even small deviations in the initial configuration will thus lead to large deviations of the resulting Bohmian trajectories. Our rapidly increasing uncertainty about the particle positions is then mirrored by the quick spreading of the wave function under the Schrödinger time evolution. Therefore, computing any one particular trajectory does not lead to useful information about the real trajectory of the system. Hence, the manifestly non-local nature of quantum mechanics is such that a system becomes immediately more chaotic as we try to minimize our ignorance regarding microscopic initial conditions. As a consequence, we have to resort to probabilistic reasoning much earlier than is often the case in classical physics. For a quantum system, Born's rule provides—provably—as good a description as we can get in a universe in quantum equilibrium.

Coming back to the comparison between Bohmian mechanics and the GRW matter density quantum theory in the preceding section, consequently, although the GRW matter density field is specified by the wave function, whereas the Bohmian particle positions are not specified by the wave function, the implications for our knowledge are similar. The reason for going for a primitive ontology theory of quantum physics is that one solves the measurement problem by endorsing a configuration of matter in physical space. But the consequence of doing so is that there is an epistemic limit to the accessibility of that configuration. Such a limited accessibility applies to any primitive ontology theory of quantum physics, independent of whether one takes the configuration of matter in physical space to consist of discrete objects (particles), as in Bohmian mechanics, or in one continuous field, as in the GRW matter density theory (see Cowan and Tumulka (2016)). In other words, one does not avoid what may seem to be a drawback of subscribing to so-called hidden variables—namely, a limited epistemic accessibility—by amending the Schrödinger equation and taking the wave function as it figures in such an amended Schrödinger equation to represent the configuration of matter in physical space. In a nutshell, if one endorses a commitment to a configuration of matter in physical space in quantum physics, one has to endorse a commitment to a limited epistemic accessibility of that configuration, in whatever way one spells out the theory of that configuration and its evolution.

By way of consequence, in any primitive ontology theory of quantum mechanics, probabilities come in through our ignorance of the exact configuration of matter in physical space. This ignorance implies that we immediately have to resort to probabilistic descriptions in quantum physics as elaborated on earlier, independently of whether the dynamical law for the evolution of the configuration of matter in physical space is deterministic (as in Bohmian mechanics) or stochastic (as in the GRW theory).

Note

1. See Belot ((2011), pp. 60–77) for a criticism that spells out how the general objections against Humeanism apply in this case.

4 A persistent particle ontology for quantum field theory

4.1 The requirements for an ontology of QFT

Any proposal for an ontology of quantum physics has to provide a solution to the measurement problem by setting out what happens in nature on the level of individual quantum systems instead of merely making statistical predictions for measurement outcomes. As regards non-relativistic quantum mechanics, we argued in Chapter 3, sections 2 and 3, that Bohmian mechanics is the most convincing proposal to solve the measurement problem. Bohmian mechanics supplements the wave equation (i.e. the Schrödinger equation) with a guiding equation that yields trajectories for individual particles in three-dimensional space (what is known as the primitive ontology) and explains the states of macroscopic systems as well as their stability in terms of these trajectories.

In doing so, Bohmian mechanics fits well with the minimalist ontology pursued in this book: the Bohmian particles are matter points without any intrinsic features. Their being individuated by their position in space means that they are individuated by the distance relations among them, which make up the configuration of matter of the universe. The wave function then is not an additional element of the physical ontology. It is a dynamical parameter employed in the laws that achieve the best combination of simplicity and information in describing the change in the distance relations among the particles (i.e. their trajectories) throughout the entire history of the universe (cf. the quantum Super-Humeanism mentioned in Chapter 2, section 3). In this chapter, we base ourselves on this minimalist ontology including Super-Humeanism applied to the dynamical parameters and laws of quantum physics, without always explicitly mentioning this background in the text and in the formalism. Our aim now is to show that the Bohmian approach does the same service for QFT that it does for quantum mechanics.

When it comes to QFT, the requirement for an ontology is the same as in quantum mechanics: the measurement problem plagues QFT in the same way as quantum mechanics (see Barrett (2014)); any proposal for its solution has to explain what happens in nature on the level of individual processes instead of merely making statistical predictions. On the one hand, a particle ontology is generally regarded as excluded for QFT, given that one of the most striking phenomena predicted by QFT is the so-called creation and annihilation of particles: if there is not a definite number of particles that persist, particles cannot be what is fundamental (see, e.g., Halvorson and Clifton (2002); Kuhlmann (2010), ch. 8; Ruetsche (2011), chs. 9–11). On the other hand, it is by no means clear that the ontology of QFT is one of fields, since QFT does not admit fields that have definite values at the points of four-dimensional space-time (see Baker (2009)). In general, if one advocates an ontology in terms of fields, the problem is to formulate a dynamics that accounts for measurement outcomes in a convincing manner, given that most measurement outcomes consist explicitly in definite positions of something—be it the tracks caused by cosmic radiation on Victor Hess's photo plates, the electron and positron tracks in Carl Anderson's cloud chamber, or the muon tracks recorded by the ATLAS detector at CERN that became famous for the Higgs-particle detection. In a nutshell, in QFT as in any other area of physics, the experimental evidence is particle evidence. Fields, waves and the like come in to explain that evidence, but are not themselves part of the evidence.

Against this background, we submit that also when it comes to QFT, the particle ontology set out in this book—matter points being individuated by the distance relations in which they stand and the change in these relations—is the simplest ontology that is empirically adequate: less will not do; bringing in more leads to drawbacks instead of providing additional explanatory value. In particular, this ontology comes with a clear and straightforward solution to the measurement problem: there always is a definite spatial configuration of matter points. In order to vindicate this ontology for QFT, we show in the following that the Bohmian approach works for QFT in the same way as for quantum mechanics: as it is a *non sequitur* to take particle trajectories to be ruled out in quantum mechanics due to the Heisenberg uncertainty relations, so it is a *non sequitur* to take permanent particles moving on definite trajectories according to a deterministic law to be ruled out in QFT due to the statistics of particle creation and annihilation phenomena. In both cases, such an underlying particle ontology is in the position to explain the statistics of measurement outcomes.

Bohm himself in his original paper Bohm ((1952b), appendix A) already discussed a possible extension of his theory to the electromagnetic field. Indeed there is a field version of Bohmian QFT (see Struyve and Westman (2006, 2007) as well as Valentini (1992), ch. 4). As regards the particle ontology, the first Bohmian QFT can be traced back to Bell (2004, ch. 19, first published 1986) (see Dürr et al. (2005) for a continuum version of Bell's proposal on a lattice; see Struyve (2010) for an overview of all the Bohmian proposals). Bell's proposal grants the entities that figure in the particle pair-creation and annihilation processes of QFT the status of real, fundamental particles, which are, accordingly, created and annihilated at random times and positions. Between these random events, the particles evolve according to a Bohmian law of motion.

We do not adopt this proposal. In the first place, since these entities depend on the choice of an initial reference vacuum state that is not unique (see Fierz and Scharf (1979)), instead of being objects that simply exist, we consider it to be inappropriate to grant them the status of fundamental objects in the ontology. More generally speaking, the Bohmian approach is motivated by finding an underlying ontology that explains the statistics of measurement outcomes. Although Bell's proposal achieves an explanation of the statistics of measurement outcomes, we think that it is worthwhile to pursue the Bohmian approach only if one is prepared to go all the way down to an ontology of fundamental objects that are simply there—that is, that do not come into existence out of nothing and that do not disappear into nothing—and that evolve according to a simple, deterministic law (that is, an evolution not interrupted by random jumps), much in the spirit of the quote from Parmenides given in Chapter 2, section 1.

That is why we take up the old idea going back to Dirac (1934): there is no particle creation or annihilation. There are only conditions under which particle motion becomes experimentally accessible or fails to be so, and these conditions are not unique; they may even depend on the state of motion of the experimental devices (as it is the case for the Unruh effect). We build an ontology of permanent particles for QFT on this idea. In this task, we face a shortcoming of the modern formulation of QFT, which is given only in terms of perturbative scattering theory. The mathematical difficulties involved in the computation of scattering matrix elements are already so severe that in the renormalization group formulation of QFT, the question about a fundamental equation of motion is not even thematized. In ontology, however, we have to make explicit what the laws are according to which individual processes occur; for what there is in nature are individual events and processes. Techniques developed to calculate measurement outcome statistics do not reveal how these processes evolve from an initial to a final state.

Despite its success, the *standard model* of QFT—as to date any other known non-trivial, Lorentz covariant, interacting classical and quantum field theory—is plagued by ill-defined equations of motion. Two central obstacles are the ultraviolet divergences of the bosonic as well as the fermionic fields. While the former type of divergence is inherited from the corresponding classical field theory of point charges, the origin of the latter can be understood best through the representation of quantum states in QFT as seas of infinitely many particles, where "infinity" would have to be conceived as a mathematical idealization, as in the thermodynamic limit of classical statistical mechanics. In such representations, quantities like the total current, then given as the sum of the respective infinitely many one-particle currents, generically diverge. It is therefore the hope that when working with relative instead of absolute quantities, these mathematical difficulties can be overcome: for instance, although the currents belonging to two quantum states may diverge individually, their difference may still be well-defined; often, the difference is already the physically relevant quantity (see for instance Scharf (1995), Deckert et al. (2010), Gravejat et al. (2013) and Mickelsson (2014) for recent mathematical rigorous attempts to arrive at a well-defined time evolution and/or vacuum polarization current). Nevertheless, except in quantum chromodynamics, in the interplay with bosonic fields, this divergence leads to the infamous Landau pole and the breakdown of renormalization theory already at a finite, though very large, energy.

These obstacles on the level of the equations of motion that persist since the 1930s notwithstanding, major progress has been made during the past 80 years on the level of scattering theory. When representing matrix elements of the scattering operator in terms of informal Taylor series in the coupling parameter of the interaction, it was shown that order by order all occurring infinities in each of the summands can be removed algorithmically by symbolic manipulation. Despite the lack of a rigorous mathematical understanding, the first orders of perturbative corrections agree astonishingly well with the experiments and speak for themselves. The higher orders are, however, untrustworthy as it is unknown whether the renormalized series of matrix elements are summable. In fact, toy models suggest that these renormalized series are asymptotic series only—that is, informal series that are divergent but may give a good approximation for small coupling parameters when only taking the first few lower order summands into account. As an example of an asymptotic series motivated by the Euclidean φ^4-theory of statistical mechanics, consider the following function

$$f(\alpha) := \int_{-\infty}^{\infty} e^{-x^2 - \alpha x^4} \, dx \tag{4.1}$$

and observe that while for any $\alpha \in \mathbb{R}$ the sequence

$$f_N(\alpha) := \int_{-\infty}^{\infty} \sum_{n=0}^{N} \frac{(-x^2 - \alpha x^4)^n}{n!} \, dx, \qquad N \in \mathbb{N}, \tag{4.2}$$

converges to $f(\alpha)$ for $N \to \infty$ with infinite radius of convergence, the sequence

$$g_N(\alpha) := \sum_{n=0}^{N} \int_{-\infty}^{\infty} \frac{(-x^2 - \alpha x^4)^n}{n!} \, dx, \qquad N \in \mathbb{N}, \tag{4.3}$$

diverges. Nevertheless, plotting $|f(\alpha) - g_N(\alpha)|$ reveals that there is an optimal $N^*(\alpha)$ such that for all $N \leq N^*(\alpha)$ the approximation gets better, while for

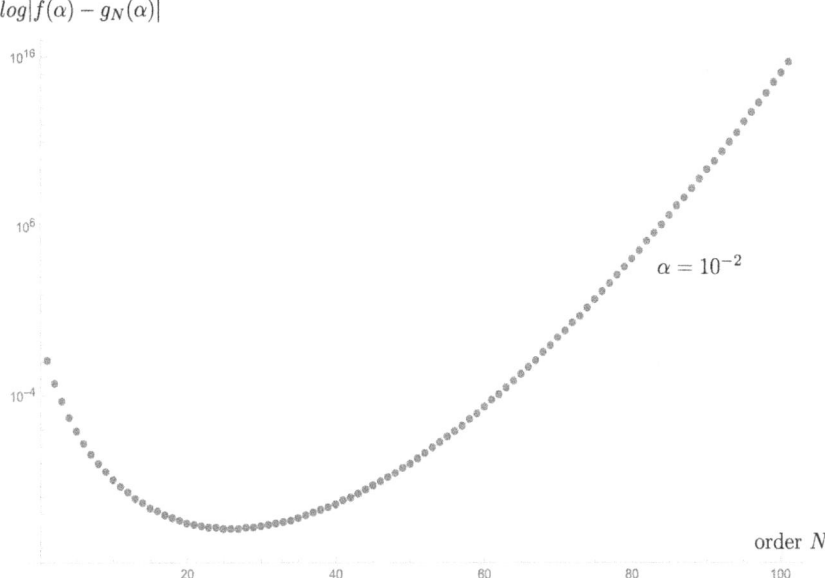

$$log|f(\alpha) - g_N(\alpha)|$$

$\alpha = 10^{-2}$

order N

Figure 4.1   This graph shows the logarithmic absolute difference between $f(\alpha)$ and its asymptotic series $g_N(\alpha)$ for the particular orders N. The optimal error lies roughly around $N^*(\alpha) \approx 22$. Until $N = N^*$ the approximation given by the asymptotic series improves steadily; for $N > N^*$ it becomes worse and the error diverges.

$N > N^*(\alpha)$ the approximation gets worse again; see Figure 4.1. Series with properties such as $g_N(\alpha)$ often arise from informal perturbation theory and are referred to as asymptotic series. In this sense, it might well turn out that next-generation experiments with sufficiently high accuracy reveal that the higher order perturbative corrections actually worsen the celebrated accuracy of the perturbative predictions of the standard model instead of improving them. The predictive power of the perturbative formulation is, hence, unclear.

The mathematical problems left aside, when it comes to the ontology of QFT, one has to be concerned with what is going on in the world according to the theory. That is to say, one has to consider the actual law of motion of the theory and not only the mapping between incoming and outgoing asymptotic particle states as it is done in scattering theory. Consequently, for the purposes of ontology, one cannot rely on the modern renormalization machinery that allows to compute finite summands of the informal and potentially divergent perturbation expansion of the scattering matrix. For our purpose, we therefore propose to return to the early attempts of the 1930s to define equations of motion for QFT. In doing so, we will of course not obtain better predictions of measurement outcomes than standard QFT. The situation is the same as in Bohmian mechanics: the guiding equation does not yield better predictions; accordingly, it does not have to be solved to calculate statistical predictions of measurement outcomes. It figures in the explanation of these statistics by answering the question of what is really going on in the world. Hence, the purpose in returning to Dirac (1934) is not to pursue outdated physics, but to make progress with respect to the ontology of QFT. In modern formulations of QFT, a mathematical description in terms of the Dirac sea, although being canonically equivalent on the level of wave functions, has been abandoned in favour of a more economic description involving particles and anti-particles as well as their creation and annihilation. It is not our intention to reintroduce the Dirac sea as computational device. On the contrary, we will argue that after it served its ontological purpose, it suggests itself to develop computational methods that let go the bulk of the Dirac sea, replacing it by a so-called vacuum state, and only track its excitations. What one gains in formulating the fundamental equations of motion in terms of the Dirac sea is an ontology that explains the appearances of particle and anti-particle creation and annihilation.

In the following, we examine the so-called *Dirac sea model* or *hole theory*, which was the first one to predict the phenomenon of electron-positron pair-creation. Although it is conceived only for the electron sector of the standard model of QFT, it is applicable to all fermionic matter. As Bell (2004, ch. 19) argued, the commitment to fermions is sufficient to account for the empirical data: they all consist in a spatial arrangement of fermions. Hence, if the Dirac sea model can cover all fermionic matter, this is all that can be reasonably demanded for it to serve as a proposal for the ontology of QFT. We should emphasize once more that the reliance on fermions being the matter points that account for the empirical data is a choice that is sufficient and suits our search for a minimalist ontology of the natural world. There are other approaches tried out in the literature: thus, Struyve and Westman (2006) consider a proposal that entirely relies on bosons. Furthermore, ontologies with a commitment to both fermions and bosons are of course also conceivable. Pursuing a minimalist ontology, however, it is suitable to choose one or the other. The simplicity by which the fermions can account for particle detection in physical space singles them out as the first and foremost candidate for the matter points in the ontology. By contrast, there is no obvious manner to account for particle detection in terms of bosons, considering, for instance, the problems connected to the position operator. Denying the bosons a place in the ontology does not make them irrelevant in our theory: they make up the dynamical structure by which the interaction between the fermions is mathematically described.

Apart from these ontological considerations, the Dirac sea model can be given a rigorous mathematical treatment in certain regimes. It is therefore suitable for probing the compatibility of QFT with an ontology of persistent particles. Colin (2003) and Colin and Struyve (2007) have carried out an investigation in that sense in the physics literature (see also the brief remarks in Bohm et al. (1987), pp. 373–374, and Bohm and Hiley (1993), ch. 12.3). In the philosophical literature, the Dirac sea model is hitherto largely ignored, apart from a brief assessment in Saunders (1999, pp. 78–79, 86–88).

In the next section, we introduce the Dirac sea model (section 4.2). Then, we define the state of equilibrium for the particles—that is, a state in which nothing can be observed, by physicists usually referred to as the "vacuum state" (section 4.3)—and small excitations from equilibrium, which can be observed and which appear as if there were a creation and annihilation of particles. This then is the basis on which we show how the formalism that effectively describes these excitations leads naturally to the standard QFT formalism, which employs creation and annihilation operators. We discuss how, due to the ontology of persistent particles, it is possible to give a clear meaning to these operators as well as to the vacuum state (sections 4.4 and 4.5). We conclude with showing how this ontology and dynamics explains the measurement outcomes, arguing that it does for QFT what Bohmian mechanics does for quantum mechanics.

4.2   The Dirac sea model

The Dirac sea model can be inferred from the standard model by imposing the following restrictions:

1.   Restriction to the electron sector of the standard model;
2.   Restriction to direct electrodynamic interaction and neglect of radiation;
3.   Modelling interaction with all other fermion sectors of the standard model only effectively by a time-dependent "external" interaction.

The resulting model is on the one hand simple enough for our discussion. On the other hand, it has sufficient structure to describe the phenomenon of electron-positron pair-creation. In terms of the particle ontology, restrictions one to three are to say that we consider only a part of the dynamical structure of the standard model—namely, only those parameters that assign to some matter points the dynamical role of being electrons (that is, move electronwise)—and neglect all other dynamical parameters. More precisely, we assume that the dynamical structure introduces a distinction of the matter points in two groups: the first group of matter points $1, \ldots, N$ for $N < M$ shall be referred to as electrons, while the rest of the matter points is assigned the role of other fermionic particles by the dynamical structure. Without restrictions one to three, the model would become unnecessarily opaque. Nonetheless, we emphasize again that the restriction to electrons is arbitrary, since the Dirac sea model and hence our argument is also applicable to the other fermion sectors and the other types of interaction in the standard model.

Electrons in direct electrodynamic interaction exclusively repel each other. If this were the only interaction, the spatial extension of this cloud of N electrons would inevitably grow larger and larger without any bound. This is an artifact of restrictions one to three, as we model only the electron sector directly—though we allow an indirect, effective interaction to other fermion sectors. To avoid this unphysical behaviour (which is due to the simplicity of the considered model) we stipulate further that, due to the motion of all the other particles and their interaction by means of an "external interaction" (which then includes also attractive interactions, maybe even gravitation, although the standard model is not yet able to describe it), the spatial extension will always be bounded (which might also be suggested by general relativity and supported by cosmological evidence). Furthermore, we assume that this "external" interaction is reasonably well behaved in the following sense: neither will it dampen the motion of the electrons to such an extent that all electron motion comes to a rest, nor will it drive the electron velocities arbitrarily close to the speed of light. Thus, there is no infinite energy transfer. The idea behind this assumption is that motion should be somewhat conserved among all fermion sectors. Neither does motion arise from nothing nor does it cease to exist, it only varies over the individual particles.

Posing these two additional restrictions allows us to avoid the discussion of mathematical problems—namely, the infamous infrared and ultraviolet divergences—which to this day prevent the formulation of a well-defined equation of motion in QFT. Put mathematically, we assume that the initial data and the "external" interaction in our theory are such that:

4. *Infrared cut-off*: The spatial extension of the universe is restricted to a finite volume Γ in 3d-space $\mathbb{R}^3$.
5. *Ultraviolet cut-off*: The electron momenta are restricted to be lower than some finite ultraviolet cut-off Λ.

In a yet to be found well-defined standard model of QFT, the earlier assumptions must not be imposed *a priori*, but have to come out *a posteriori*. In particular, the gained mathematical well-definedness of these *ad hoc* stipulations comes at the cost of a violation of Lorentz invariance and the introduction of seemingly arbitrary parameters. Accordingly, the main objective of modern renormalization theory is the removal of these unwanted cut-offs. As yet, however, a removal is only achieved order by order in the informal (and as discussed most likely asymptotic) expansion series of the scattering matrix elements. When it comes to defining an equation of motion, the introduction of the aforementioned cut-offs is to date unavoidable in QFT. The ultraviolet cut-offs allow for a self-adjoint interaction Hamiltonian in the corresponding second-quantized Schrödinger equation, while the infrared cut-off is for example needed in order to find Fock space representations of the far-field of massless particles (due to a gauge symmetry violation, all the particles except for the Higgs boson are at first assumed to be massless—the masses are later generated by the spontaneous symmetry breaking induced by the Higgs mechanism). Nevertheless, it is the well-founded hope amongst physicists that, provided the infrared and ultraviolett cut-offs are, respectively, small and large enough, and carrying out an appropriate renormalization scheme, the effects on the dynamics are only small in certain regimes of interest. In fact, given the cut-offs, the formulation presented here is equivalent to the textbook formulation on the level of wave functions. Furthermore, it is possible to include all gauge fields of electroweak interaction to allow for a description of radiation effects in order to remove assumptions one to three altogether (see Colin and Struyve (2007)). In general, the Dirac sea model applies to any fermion sector and potentially any interaction between those fermions (except maybe gravitation whose quantum nature is unclear). In short, the earlier assumptions are no objectionable limitations of our endeavour. They rather underline the general point that the rationale of the Bohmian approach is not to obtain better physics, but an ontology for the existing physics.

The Dirac sea model was originally proposed by Dirac (1934, 1947) in Hartree-Fock approximation in terms of density matrices. Without this approximation it was later included in QFT by means of a second quantization procedure of the one-particle Dirac equation yielding the so-called Dirac field operators on Fock space. We discuss the Dirac sea model in terms of a straightforward quantum-mechanical, N-particle wave equation, a formulation which, for finite Γ and Λ, is equivalent to the textbook version by means of second quantization of the one-particle Dirac equation with Coulomb pair interaction (see section 4.4). Additionally, we conceive the Dirac sea model in terms of Bohmian mechanics.

The Bohmian velocity field v_t guiding the motion of N particles can be given in the form

$$v_t(X) = c(j_t^{(k)}(X)/\rho_t(X))_{k=1,\ldots,N}, \qquad (4.4)$$

$$\rho_t(X) = \Psi_t(X)^*\Psi_t(X), \quad j_t^{(k)}(X) = \Psi_t(X)^* 1^{\otimes(k-1)} \otimes \alpha \otimes 1^{\otimes(N-k)}\Psi_t(X), \quad (4.5)$$

where $X = (x_1,\ldots,x_N) \in \Gamma^N$, and ρ_t and $j_t^{(k)}$ for $k = 1,\ldots,N$ are the probability density and currents generated by a wave function Ψ_t and c stands for the speed of light. Here, Ψ_t denotes an anti-symmetric, square-integrable, N-particle spinor-valued function on configuration space $\Gamma \subset \mathbb{R}^{3N}$, in short $\Psi_t \in \mathcal{H}^{\wedge N}$ for $\mathcal{H} = L^2(\Gamma, \mathbb{C}^4)$, that solves the wave equation

$$i\hbar\partial_t\Psi_t(x_1,\ldots,x_N) = H^N\Psi_t(x_1,\ldots,x_N),$$

$$H^N = \sum_{k=1}^{N}(H_k^0(x_k) + V_k(t,x_k) + H_k^I(x_k)). \qquad (4.6)$$

The Hamiltonian H^N is made of the free Dirac Hamiltonian $H_0^k(x)$; the external influences $V_k(x)$ are given by

$$H_k^0(x) = 1^{\otimes(k-1)} \otimes H^0(x) \otimes 1^{\otimes(N-k)}, \qquad H^0(x) = -ic\alpha \cdot \nabla_x + \beta mc^2,$$

$$V_k(t,x) = 1^{\otimes(k-1)} \otimes V(t,x) \otimes 1^{\otimes(N-k)}, \qquad \text{for some external potential } V(t,x),$$

(4.7)

where m stands for the electron mass and $\otimes$ denotes the tensor product (note that Ψ is spinor-valued). Furthermore, β and the components of the vector $\alpha = (\alpha_1, \alpha_2, \alpha_3)$ are $\mathbb{C}^{4 \times 4}$ matrices that fulfill the anti-commutator relations $\{\alpha_j, \alpha_k\} = 0 = \{\beta, \alpha_j\}$ and $\beta^2 = 1 = \alpha_j^2$ for $j \neq k$, $j,k \in \{1,2,3\}$. Moreover, H_I denotes the Hamiltonian modelling the interaction between the particles. Since we neglect radiation, the interaction is mediated directly by

$$H_k^I(x) = \frac{1}{2} \sum_{j \neq k} U(x - x_j),$$

(4.8)

where the electric interaction between the N electrons can be taken to be the Coulomb potential $U(x) = \frac{e^2}{4\pi\epsilon_0} |x|^{-1}$. Here, ϵ_0 is the electric constant and $e < 0$ the charge of an electron. Note that H_k^I depends also on x_j, which is suppressed in our notation. Finally, the composite, effective interaction of all particles on electron k is given through a time-dependent potential $V_k(t,x)$, stemming for example from the Coulomb field of a present ion, etc.

In sum, the velocity field (4.4) and the wave equation (4.6) define what we call the Dirac sea model—an N-particle version of Bohmian mechanics with Dirac's instead of Schrödinger's dispersion relation. Given an initial configuration Q_0 and wave function Ψ_0, the corresponding solutions of these equations yield a unique trajectory of configurations of the N electrons parametrized by time. In principle, there is nothing more to say about the ontology: there are N persistent matter points that move electron-wise, that is, evolve according to the deterministic law of motion given by (4.4)–(4.6). However, we face the following situation:

I.   Generically, we do not have complete knowledge about the initial data (Q_0, Ψ_0).
II. Even if we did, if N is large (which is the interesting case, as we assume large cut-offs), it is in general neither analytically nor numerically feasible to compute solutions Ψ_t of the wave equation (4.6).

Problem (I) is generic to any theory that applies to the universe as a whole. Predictions about subsystems have to be inferred from a detailed statistical analysis of what is to be expected in most situations with respect to a meaningful measure. In Bohmian mechanics, this is made possible by the special form of the velocity law (4.4), which ensures that if the initial data Q_0 is distributed at random according to $|\Psi_0(X)|^2 d^{3N}X$, configuration Q_t is distributed according to $|\Psi_t(X)|^2 d^{3N}X$. This feature is known as equivariance. The statistical analysis of subsystems has been carried out in Dürr et al. (2013b, ch. 2) for Bohmian mechanics. We discussed it in Chapter 3, section 4. This analysis applies here as well. Its bottom line is a proof that Born's rule holds true: if the effective wave function of the subsystem is given by $\varphi_t(x_1,. . .,x_n)$ for $n \leq N$, then the distribution of the subconfiguration of particles $q(t) = (q_1(t),. . .,q_n(t))$ is given by $|\varphi_t(x)|^2 d^n x$.

Problem (II) stems from the fact that the pair potential U strongly entangles all tensor components of the wave function Ψ_t during the time evolution. Even a perfect initial anti-symmetric product state will therefore immediately lose its product structure due to (4.6). The complexity of this entanglement increases exponentially with N. Even today's super computers fail to solve the Schrödinger equation in general situations in three spatial dimensions for more than three particles. This is already due to memory requirements, not to mention the necessary computational power. Consequently, one must hope to find special non-trivial situations in which the very complicated N-particle dynamics becomes simple enough so that it can be approximated by studying the motion of a few particles only.

The original motivation for Dirac's hole theory was not the complexity of its solutions, but stemmed from the attempt to make sense of Dirac's equation describing only one particle (see Saunders (1991) for the context of Dirac's theory). One quickly realized that the solutions showed strange behaviour—for instance the *Zitterbewegung* (Schrödinger (1930)) and Klein's paradox (Klein (1929)—which made Dirac's equation hard to interpret physically. The mathematical reason for this strange behaviour is the presence of a negative energy spectrum of the free Dirac Hamiltonian $H^0(x)$ as given in (4.7). Therefore, the starting point of Dirac's hole theory was to somehow suppress this negative spectrum. As the electrons move like fermions, this can be accomplished by filling all the negative energy states with a sea of particles—and, granted the cut-offs four to five, thus ending up with an N-particle wave equation as noted earlier. The Pauli exclusion principle then prevents the wave function of any additional electron from growing negative energy components, because all negative energy states are already occupied. The consequence is of course that one ends up with the same complexity problem (II) as well, albeit for a different reason.

Our starting point is different. We seek a theory about the universe as a whole, based on an ontology of permanent particles. We bring in Dirac's hole theory as a suitable means to implement that ontology. This model together with its ontology is spelled out concisely in this and the preceding section. However, even though we only treat a simplified model of that theory and only describe electrons while modeling the rest of the particle interactions effectively, it is natural to start with not only one but all the electrons of the universe. The model can thus only be considered as a serious contender, if it is also possible to analyze it mathematically despite of the complexity problem (II), which is why we anyhow have to cope with a large number of electrons N. This is the content of the rest of this chapter of the book: our aim is to show how on the basis of this model we can explain the statistical predictions of standard QFT.

Before doing so, let us compare our approach with the discussion of the Dirac sea model in a Bohmian framework by Colin and Struyve (2007). They also assume a finite fermion number; they are committed to positions for fermions for both positive and negative energy particles, whereas there are no bosons in the ontology. Anti-particles are defined as holes in the sea of the negative energy particles. Whereas Colin and Struyve (2007) define the fermion number as the total number of particles minus the number of negative energy particles plus an infinite constant, we are committed to N matter points that de facto coincide with the fermions. Nonetheless, even though in our model there is no ontologically significant difference between positive and negative energy particles, one may say with the usual jargon that the fermion number can be defined as the number of positive energy particles plus the number of negative energy particles, which remains constant. Colin and Struyve (2007) propose an equation of motion for fermions that defines the vector velocity field for configurations of particles, which is dependent on the wave function of the system according to the usual Bohmian recipe; the expectation value of the fermion number density is related to the position density: intuitively, the number of fermions in a given region of space corresponds to the fermion positions in that region. This model is regularized with the introduction of ultraviolet momentum cut-offs and finite space; these constraints ensure that it is a mathematically well-defined theory. Equivariance guarantees that the empirical predictions of the standard model are reproduced, exactly as in our case.

We emphasize again that so far the standard model of QFT only permits to remove the introduced cut-offs when computing informal perturbative corrections of scattering amplitudes with respect to the non-interacting QFT, while the respective equations of motion for the standard model become ill-defined with any attempt to remove the cut-offs. Since both the theories of Colin and Struyve (2007) and the one developed in this book are concerned with the dynamics and not with scattering theory, they can only be compared to the standard model with cut-offs—until eventually a formulation of the equations of motion of the standard model without cut-offs is found. Nevertheless, this does not infringe upon the experimental adequacy of the models with cut-offs. On the contrary, there is a vast spectrum of experimental evidence, starting from the historical Lamb shift calculation and ranging to recent results, such as the independence of the effective particle masses of the infrared cut-off. In short, these models agree with the experiments in large regimes, provided the cut-offs are sufficiently large and an appropriate renormalization scheme is applied.

4.3   Equilibrium states and the vacuum

How can we tackle problem (II)—that is, find manageable approximations of the in general very complicated dynamics of the Dirac sea model? Clearly, this will be possible only in special situations—that is, for a certain class of initial quantum states. Furthermore, such an approximation cannot be carried out without coarse-graining the level of detail of the information that is to be inferred about the system.

Consider, as an analogy, a classical gas of N particles confined to the volume Γ. Although it is in principle possible to infer the actual motion of the individual particles by solving Newton's law of motion, for large N, this would be a hopeless endeavour and for many practical purposes unnecessary. For instance, even in equilibrium, the *microscopic*, actual Newtonian motion of the particles may be very intricate. However, effectively the net result is that, *macroscopically*, the gas density is almost constant with the variation around this constant density being small. The smallness requirement depends on the practical purpose as it determines which effects will be visible and which ones will drown in the fluctuations. Hence, it defines what is meant by *macroscopic*. In the same vein, one can introduce further macroscopic parameters like volume, pressure and temperature. For most engineering purposes their relationship constitutes a satisfactory description of equilibrated gases—that is, the theory of thermodynamics. Such a coarse grained description is not restricted to equilibrium states only. For example, in certain regimes it is possible to describe the mediation of a perturbation created in the gas by an external influence in terms of an effective equation for pressure (sound) waves that excite the initially equilibrated gas, without knowledge of the actual microscopic Newtonian motion of the individual particles. Using this classical example as a guide, the idea of describing complex dynamics in terms of small deviations from an equilibrium whose time evolution can be understood in more simple terms will be the key ingredient.

In quantum mechanics, much progress has been made in deriving such coarse-grained or approximate descriptions with full mathematical rigour in the case of, for instance, Bose-Einstein condensates and large molecules (for recent works on Bose and Fermi gases see, e.g., Pickl (2011), Benedikter et al. (2014)). The strategy underlying this type of analysis is to show that in certain regimes—such as large particle number N, high or low density, etc.—the reduced one-particle density matrix of the N-particle system described by Ψ_t can be well approximated by the one of a product state, although the actual wave function Ψ_t is far from being a product state. The product state has a much simpler time evolution. This comes at the cost of the level of detail that can be inferred from such an approximation, which is then restricted to questions that can be answered by knowing the reduced density matrices only. In many situations this is, however, all that is needed. The mechanism that is usually exploited to derive such approximations rigorously is that if the variation of the interaction operator—e.g. such as H_k^I in (4.6)—is sufficiently small, then it can be replaced by its expectation value for a certain time interval without accumulating too much of an error. In the following, we set up a similar strategy for the N-particle system of the Dirac sea, although we have to concede that the contemporary mathematical techniques do not yet allow us to control the approximation made in full rigour.

In this respect, we depart from the approach of Colin (2003) and Colin and Struyve (2007). They, too, analyze the motion of the N particles in the Dirac sea by comparing it to the motion that the particles would conduct in a special state—namely, the one physicists call the vacuum state. Colin and Struyve (2007) argue in the spirit of Dirac (1934) that, as the latter gives rise to on average very uniform distributions of particles, deviations from uniformity are observable provided the fluctuations around the uniform motion in the vacuum state are sufficiently small. The distinction of the observable motion against that uniform background is then based on the expected particle densities and their fluctuations, for which explicit estimates are given. We on the contrary look directly for approximate solutions to (4.6). In doing this, we similarly track the motion of the Dirac sea of N particles in terms of deviation from the motion of a reference state of the Dirac sea. However, we do so already on the level of wave functions. This will then naturally lead to the standard QFT formalism in terms of creation and annihilation operators. Furthermore, we argue that the choice of reference state is not unique but rather a matter of choice of "good coordinates" for a representation of states with respect to which one can economically arrive at approximate solutions to (4.6).

Let us pursue the idea of describing complex dynamics in terms of small deviations from an equilibrium whose time evolution can be understood in more simple terms. At first, we only consider $V(t,x)$ in equation (4.6) to be zero. That is to say, we stipulate that there are no external influences: the motion of the electrons can be represented in terms of (4.4) and (4.6)—that is, subject to the Fermi repulsion (due to the anti-symmetry of Ψ_t) and the Coulomb repulsion only. In analogy to the classical gas, the simple approximate solutions inferred in the following play the role of the equilibrium states. In the next section, we then consider the case $V(t, x) \neq 0$, which leads to excitations of the equilibrium—the analogue to the pressure waves in the classical example.

As already indicated, the term creating problem (II) is the Coulomb pair-interactions encoded in H_k^I given by (4.8). Although the general dynamics may be very complicated, there may be special initial conditions that lead to a particularly simple dynamics, and thus, form a good reference system to study more complicated dynamics. We therefore define a class of initial quantum states that allow an approximation in which this interaction term effectively vanishes and call it the class of equilibrium states—that is, our reference states. This class shall consist of all states $\Psi \in \mathcal{H}^N$ that fulfill the following two conditions: (a) The quantum mechanical expectation of the interaction operator H_k^I is approximately constant—that is, there is a constant E^I such that the average interaction approximately fulfills

$$E^I \approx \langle \Psi, H_k^I(x)\Psi \rangle = \left\langle \Psi, \frac{1}{2}\sum_{j\neq k} U(x - x_j)\Psi \right\rangle \quad \text{for all } x \in \mathbb{R}^3 \text{ and } k = 1, ..., N, \quad (4.9)$$

and (b) the fluctuation around this constant expectation value is sufficiently well behaved throughout the time evolution. Condition (a) states that the pair-interaction between the particles effectively averages out initially, while condition (b) implies that this feature is preserved over time.

In view of the weak law of large numbers, these conditions (a) and (b) can only be met for sufficiently large N. As the theory so far is meant to apply to the total number of electrons in the universe, N is naturally large. The exact sense of "approximatively" and "sufficiently" depends on the practical purpose: conditions (a) and (b) are there to make sure that the solutions to the fundamental equation of motion (4.6) for $V(t,x)$ = 0, given an initial state Ψ_0 in this class of equilibrium states, can for all practical purposes be sufficiently well approximated—e.g. in the sense of reduced density matrices—by a solution to the much simpler effective equation of motion

$$i\hbar\partial_t \Psi_t^{\approx}(x_1, ..., x_N) = \sum_{k=1}^N \left(H_k^0(x_k) + E^I\right)\Psi_t^{\approx}(x_1, ..., x_N), \quad (4.10)$$

replacing the complicated interaction H_k^I by the constant E^I.

Regarding the precise mathematical requirements of conditions (a) and (b) and the precise sense of the approximation we are purposely vague, since the exact behaviour of the fluctuation needed to carry out the rigorous mathematics is not entirely settled in the fermionic case (unlike the bosonic case, where for example Gross-Pitaevski and mean-field approximations can be rigorously derived). If an initial state Ψ_0 fulfills (a) and (b) in a sufficiently strong sense, the corresponding fully interacting time evolution (4.6) is close to the non-interacting one (4.10) as the errors which depend on the accumulated fluctuations (b) around the expectation value (a) can be controlled—at least in an appropriate sense, such as the one of reduced density matrices and for large enough N and gas densities. However, while fermionic N-particle states that fulfill (a) are known (e.g. the non-interacting fermionic ground state, see (4.15)), it is unknown whether there are ones that fulfill also condition (b) in a sufficiently strong sense. In fact, it is currently conjectured that (b) might be too strong and that demanding sufficiently small fluctuations for only those particles with momenta below some threshold might already suffice. This would mean that condition (b) could even be weakened and we would still be able to show the closeness of (4.6) to the approximate time evolution (4.10) for such states. The exact notion is however irrelevant to our discussion. The important point is that (4.10) does not have to be assumed, which would be highly questionable for an interacting Fermi gas, but can be derived from much simpler and plausible conditions such as (a) and (b).

Before considering a pertinent example of such an equilibrium state, let us put the motivation for focussing on the class of equilibrium states in other terms. Assuming that there are only electrons subject to Fermi and Coulomb repulsion, the ground state Ψ^{gs} of such an electron gas is expected to be one in which for almost all initial Q_0 with respect to the relevant measure $|\Psi^{\mathrm{gs}}(X)|^2 d^{3N}X$, the electrons are very homogeneously distributed. Then, if the measure $|\Psi^{\mathrm{gs}}(Q)|^2 d^{3N}X$ gives rise to a homogeneous distribution, our defining condition (4.9) of the class of equilibrium states is a consequence, and the net effect of H_k^I is a constant potential, canceling out all interactions: on the level of wave functions, the electrons in such a state effectively do not "take notice" of each others presence. Effectively, they move as if they were in a vacuum. Hence, the dynamics generated by (4.6) is very simple, and, within the bounds of the approximation, it leaves the class of equilibrium states invariant.

A natural representative for a state in the equilibrium class would be the actual ground state Ψ^{gs} of the interacting system. Its mathematical structure, however, is extremely complicated and to date not accessible due to the discussed entanglement induced by the H_k^I terms. Consequently, we have to find a simpler candidate that replaces Ψ^{gs} in the sense of the aforementioned approximation. Physicists usually choose the ground state $\Psi_{\approx}^{\mathrm{gs}}$ of the corresponding approximate equation (4.10) given by

$$\sum_{k=1}^{N} H_k^0(x_k)\,\Psi_{\approx}^{\mathrm{gs}} = E_{\approx}^{\mathrm{gs}}\,\Psi_{\approx}^{\mathrm{gs}}, \tag{4.11}$$

where $E_{\approx}^{\mathrm{gs}}$ is the lowest eigenvalue. Since the $H_k^0(x_k)$ commute pairwise, $\Psi_{\approx}^{\mathrm{gs}}$ can be found by studying the spectrum of $H^0(x)$ only.

Due to the finite volume Γ and momentum cut-off Λ, the momenta are restricted to $p \in \mathcal{P}_{\Gamma}^{\Lambda} = \{k = (k_1, k_2, k_3)\mid \parallel k \parallel \le \Lambda, k_i = 2\pi/Ln_i, n_i \in \mathbb{Z}, i = 1, 2, 3\}$—in the case of Γ being a cube of length $L = \Gamma^{1/3}$—; for each admissible momentum value there are four one-particle eigenstates

$$H_0 \epsilon_{p,\uparrow}^+ = +E_p \epsilon_{p,\uparrow}^+, \quad H_0 \epsilon_{p,\downarrow}^+ = +E_p \epsilon_{p,\downarrow}^+, \quad H_0 \epsilon_{p,\uparrow}^- = -E_p \epsilon_{p,\uparrow}^-, \quad H_0 \epsilon_{p,\downarrow}^- = -E_p \epsilon_{p,\downarrow}^-. \tag{4.12}$$

These are characterized by the sign of the eigenvalue whose modulus is determined by $E_p = \sqrt{p^2 + m^2}$ and by a mathematical entity referred to as *spin*, which is denoted by the subscript $\uparrow$ or $\downarrow$. The one-particle Hilbert space $\mathcal{H} = L^2(\Gamma, \mathbb{C}^4)$ can then be split into two subspaces spanned by the eigenstates of positive and negative eigenvalues, denoted by $\mathcal{H}^+$ and $\mathcal{H}^-$

respectively. The ground state of the eigenvalue equation (4.11) can subsequently be given by the anti-symmetric tensor product

$$\Psi_{\approx}^{gs} = \varphi_1 \wedge \varphi_2 \wedge ... \wedge \varphi_N, \tag{4.13}$$

where $\varphi_n \in \{\epsilon_{k,\kappa}^{\sigma} \mid \sigma = \pm, k \in \mathcal{P}_{\Gamma}^{\Lambda}, \kappa = \uparrow, \downarrow\}$, $n = 1, 2, . . ., N$, are eigenstates (4.12) with lowest possible eigenvalues. Due to the anti-symmetric tensor product $\wedge$, these states must be pairwise distinct for Ψ^{gs} to be non-zero; however, their particular enumeration with respect to n does not play a role. The total ground state energy $E_{\approx}^{gs}$ given in (4.11) equals the sum of the corresponding eigenvalues.

Among all such N-dependent ground states Ψ^{gs}, the one for $N = 2|\mathcal{P}_{\Gamma}^{\Lambda}|$, in which the total number of matter points equals twice the number of admissible momenta in the set $\mathcal{P}_{\Gamma}^{\Lambda}$ (since there are two spin values), is distinguished. In this case, $\Psi_{\approx}^{gs}$ is built solely out of all eigenstates with negative eigenvalues. Adding another eigenstate, which then must have a positive eigenvalue, increases the total ground state energy $E_{\approx}^{gs}$. Taking one eigenstate away also increases the ground state eigenvalue $E_{\approx}^{gs}$, since the corresponding negative eigenvalue must be subtracted. Because of this fact, one usually considers the case of a sea of $N = 2|\mathcal{P}_{\Gamma}^{\Lambda}|$ many particles and denotes the corresponding ground state $\Psi_{\approx}^{gs}$ by $\Omega := \bigwedge_{k \in \mathcal{P}_{\Gamma}^{\Lambda}, \sigma = \uparrow, \downarrow} \epsilon_{k,\sigma}^{-}$, which in physics is referred to as the Dirac sea of the *vacuum* as previously mentioned. Note that this is only a convenient mathematical idealization (and might not be fully obeyed in nature as we seem to see slightly more matter than anti-matter). Due to the anti-symmetric tensor product $\wedge$ one may also choose $(\varphi_n)_{n = 1, . . ., N}$ to be any other orthonormal basis of the subspace $\mathcal{H}^- \subset \mathcal{H}$. This changes the definition of Ω at most by a constant phase factor, which is irrelevant to the dynamics. Therefore, neglecting the phase factor, we will represent Ω in the future as

$$\Omega = \varphi_1 \wedge \varphi_2 \wedge ... \wedge \varphi_N \tag{4.14}$$

for an arbitrary orthogonal basis $(\varphi_n)_{n=1, . . ., N}$ of $\mathcal{H}^-$.

It can now be checked directly that for $\Psi = \Omega$ in the equilibrium condition (4.9), one gets

$$(4.9) = \langle \Omega, H_k^I(x)\Omega \rangle = \frac{1}{2}(N-1)\frac{1}{N}\sum_{n=1}^{N} \int U(x-y)|\varphi_n|^2(y)\, d^3y, \tag{4.15}$$

which is constant (note that $|\varphi_j| = 1/|\Gamma|^{1/2}$). This agrees with Dirac's heuristic picture that, in an equilibrium state, the electrons are uniformly distributed as, according to Born's rule, the right-hand side of (4.15) can be interpreted as expectation value of $\frac{1}{2}\sum_{j\neq k} U(x-x_j)$ for $(N-1)$ uniformly distributed random variables x_j. Beside this, there are unfortunately only few rigorous results on the quality of the approximation of (4.10) to (4.6)—that is, whether Ω also fulfills condition (b) in an appropriate sense to qualify as a state in equilibrium. Nevertheless, the indication that it holds in certain physically interesting regimes comes from the overwhelming accuracy achieved by predictions obtained by means of using (4.10) whose justification requires conditions (a) and (b) as *a priori* assumptions.

Before continuing our investigation of Ω, let us once more come back to the discussion of the introduced cut-offs four and five. It is a mathematical fact that when removing one of them (or both), the number of states in the negative spectral subspace $\mathcal{H}^-$—that is $2|\mathcal{P}^\Lambda_\Gamma|$, becomes countable infinite. Hence, filling completely $\mathcal{H}^-$ is at tension with a primitive ontology of finitely many persistent particles. It is therefore no option for our approach. Since we have only a finite number of particles in the ontology, we can only fill finitely many states in $\mathcal{H}^-$, leaving the remaining infinitely many ones unoccupied.

One may therefore raise the following concern: if one couples these particles to an open dynamical system such as the Maxwell field without further constraints, an electron (even deep down in the sea) may pump energy into the Maxwell field degrees of freedom by means of radiation and thus acquire an ever more negative kinetic energy state. Since without cut-offs there is no lower bound in $\mathcal{H}^-$, this process may go on forever so that it can eventually cause the entire dynamical system to become unstable; this is the so-called radiation catastrophe (see Greiner and Bromley (2000)). However, an open system with an energy sink at spatial infinity into which the radiated energy may simply disappear without ever interacting again with the sources is of course no viable candidate for a fundamental theory (not to mention the effects of the escaping energy in a general relativistic setting). Hence, on top of the mathematical difficulties discussed earlier, removing the cut-offs must be done in a physically sensible way in order to avoid a dynamic instability.

One way to achieve this is to conceive the interaction between the fermionic fields and the boson fields as requiring that all radiation that has been emitted by a source at a time instant has to be absorbed again by some other source at another time instant. Such a requirement removes by definition the discussed dynamic instability. If an electron were to radiate and thereby to acquire a more negative kinetic energy state, another electron would have to absorb the radiated energy and acquire a more positive kinetic energy state. However, the larger the energy gap between the electrons is, the less likely is an interaction between them. This means that depending on the initial state of the system, there are effective upper and lower bounds on the energy of the electrons so that the system is dynamically stable.

Nonetheless, until today, the mathematically status of such a requirement is unknown, not to mention a proof of the dynamic stability. That notwithstanding, the given argument is a sound indication that one does not have to abandon an ontology of finitely many particles when removing the cut-offs. Furthermore, as a mathematical idealization, it is of course completely admissible to pass over to the thermodynamic limit (as done in classical statistical mechanics), in the sense of infinitely many particles and an infinite volume at constant density, whenever the corresponding dynamical structure thereby becomes mathematically more tractable. As in classical statistical mechanics, however, this limit has no bearing on an ontology of finitely many persistent particles.

In Bohmian mechanics, all there is to the particles comes down to their position in space only and the evolution of this position as given by a guiding equation, in our case equation (4.4). In other words, in Bohmian mechanics, particles are primitive objects in the sense that they do not have intrinsic features, but are individuated only by the relative distances among them. All the other parameters including mass, charge, spin, etc., are not additional elements of the ontology characterizing the particles, but dynamical parameters employed to describe the evolution of the relative distances among the particles. Against this background, we emphasize again that an energy value is nothing but a dynamical parameter capturing a particular form of motion. Hence, admitting negative energy values poses no problem in an ontology that is committed only to particles as characterized by their position in space—in other words, matter points as individuated by distance relations and their change. Negative energy values then are only a bookmark to keep track of a certain evolution of these distance relations. The same goes for positrons and anti-matter in general: all these are matter points moving in a certain manner.

In sum, we have defined a class of initial quantum states—namely, the equilibrium class—that allow us to solve the, in general, very complicated dynamics (4.6) approximately by the much simpler dynamics (4.10). Furthermore, we have found a particularly simple representative Ω of this class. Solving the approximate equation (4.10) for the initial value $\Psi_0^{\approx} = \Omega$ yields the explicit and particular simple evolution $\Psi_t^{\approx} = \Omega_t$ for

$$\Omega_t = e^{-i(E_{\approx}^{gs} + NE^l)t}\Omega, \tag{4.16}$$

which can now be taken as approximate solution Ψ_t to equation (4.6) for the same initial value $\Psi_0 = \Omega$. Such a state only gives rise to a trivial dynamics of a sea of N electrons in which none of them "takes notice" of the rest as if they were in a "vacuum"—hence the name *vacuum state*. Nevertheless, this state Ω will provide the basis for studying more interesting dynamics in the next section.

Finally, it has to be emphasized that even with Ψ_t and Ω_t being close initially for $\Psi_0 = \Omega$ in the sense of, for instance, reduced density matrices, little information is provided about the closeness of the velocity fields (4.4) generated by Ψ_t and $\Psi_t^{\approx}$, respectively. In general, the respective velocity fields and the corresponding trajectories differ. However, the results of their statistical analysis agree due to the mode of the approximate agreement in terms of, for instance, the reduced density matrices. In other words, the price that one has to pay to overcome the complexity problem (II) by means of an approximation in terms of solvable equations is that one has to abandon the hope of obtaining a calculation of actual trajectories in favour of a statistical analysis. This is, however, the same situation as in the example of the classical gas discussed at the beginning of this section. Furthermore, it is the same situation as in Bohmian mechanics: there also is no point in calculating individual trajectories, since tiny deviations in the initial configuration will lead to large deviations of the resulting Bohmian trajectories. Consequently, our knowledge of subsystems of the universe is limited to what can be obtained from Born's rule, as explained in Chapter 3, section 4.

4.4 Excitations of the vacuum and the Fock space formalism

More interesting dynamics will take place if we allow $V(t, x)$ to be non-zero, thus including external influences. Under the action of $V(t, x)$, an initial vacuum state Ω, as defined in (4.14), may evolve into an excited state, as for instance

$$\Phi = \chi \wedge \varphi_2 \wedge \dots \wedge \varphi_N, \tag{4.17}$$

where the element φ_1 of the orthonormal basis of $\mathcal{H}^-$ is replaced by a wave function $\chi \in \mathcal{H}^+$, a so-called electron-position pair. For the sake of simplicity, let us assume that after the transition from Ω to Φ, the external influence vanishes again. As in the preceding section we strive to find an economic effective description of the time evolution of Φ to a Φ_t, $t > 0$ that fulfills the complicated dynamics (4.6)—that is,

$$i\partial_t \Phi_t = H^N \Phi_t, \tag{4.18}$$

where H^N is the N-particle Hamiltonian defined in (4.6).

In analogy to the pressure waves in the classical gas example mentioned in the previous section, the leading idea is to describe the complex evolution of Φ_t by the evolution of the excitation only as compared to the simple evolution of the reference state Ω_t given in (4.16). In mathematical terms this can be done by the following ansatz

$$\Phi_t^\approx (x_1, \dots, x_N) = \sum_{k=1}^{N} (-1)^{k+1} \int d^3 y_k\, \eta_t(x_k, y_k)\Omega_t(x_1, \dots, y_k, \dots, x_N). \tag{4.19}$$

The summation and the factor $(-1)^{k+1}$ are to ensure the anti-symmetry of the wave function. For the choice

$$\Omega_{t=0} = \Omega, \qquad \eta_{t=0} = \chi \otimes \varphi_1^*, \tag{4.20}$$

one obtains $\Phi^{\approx}_{t=0} = \Phi$ due to the orthonormality of the states χ, φ_1, . . .,φ_N. The excitation is hence encoded by a two-particle wave function $\eta_t(x, y)$. Its x tensor component, which we shall refer to as *electron component*, tracks the evolution of the initial excitation χ. Its y tensor component, which we shall refer to as *hole component*, tracks the evolution of the corresponding state in the reference Ω_t that is missing.

As these components might entangle during the evolution, they cannot be described by separate one-particle wave functions. In order to find a good effective evolution equation for η_t, one has to study the evolution of the difference $\Phi_t - \Phi^{\approx}_t$, which due to the unitarity of the time evolution amounts to gain good control of

$$\frac{d}{dt}\Phi_t, \Phi^{\approx}_t = -\langle iH^N\Phi_t, \Phi^{\approx}_t\rangle \tag{4.21}$$

$$+\left\langle \Phi_t, \sum_{k=1}^{N}(-1)^{k+1}\int dy_k\left(\frac{d}{dt}\eta_t(x_k, y_k)\right)\Omega_t(x_1, ..., y_k, ..., x_N)\right\rangle \tag{4.22}$$

$$+\left\langle \Phi_t, \sum_{k=1}^{N}(-1)^{k+1}\int dy_k\eta_t(x_k, y_k)(-iH^N)\Omega_t(x_1, ..., y_k, ..., x_N)\right\rangle. \tag{4.23}$$

We observe that

$$\int dy_k\left(H^N\eta_t(x_k, y_k)\Omega_t(x_1, ..., y_k, ..., x_N) - \eta_t(x_k, y_k)H^N\Omega_t(x_1, ..., y_k, ..., x_N)\right) \tag{4.24}$$

$$= \int dy_k\left((H^0_k(x_k)\eta_t(x_k, y_k) - \eta_t(x_k, y_k)\overline{H^0_k}(y_k) - U(x_k - y_k)\eta_t(x_k, y_k)\right) \tag{4.25}$$

$$+\left(\sum_{\substack{m=1\\m\neq k}}^{N}U(x_k - x_m) + U(x_k - y_k) - \sum_{\substack{m=1\\m\neq k}}^{N}U(y_k - x_m))\eta_t(x_k, y_k)))\Omega_t(x_1, ..., y_k, ..., x_N\right) \tag{4.26}$$

holds, where $\overline{H^0}(y) = -i\alpha \cdot \nabla_y + \beta m$ with the gradient acting to the left. The terms in (4.26) have two roles. On the one hand, they describe how, due to the presence of the excitation η_t, the excited Dirac sea deviates from the equilibrium condition (4.9)—a phenomenon that is referred to as *vacuum polarization*. On the other hand, they describe the back reaction on the electron and hole components of η_t. Note the relative signs: the first two terms in (4.26) exert a repulsive interaction on the electron component of η_t stemming from the $x_1, \ldots, y_k, \ldots, x_N$ components of the Dirac sea, while the third one exerts an attractive interaction on the hole component of η_t due to the x_n, $n \neq k$, components of the Dirac sea—exactly as one would expect, since the hole component describes the absence of one electron in the sea. Hence, it gives rise to the same charge of an electron, however, with opposite sign. According to Born's rule, which applies thanks to the discussed mode of approximation, the position distribution of these two charges is given by $|\eta_t(x, y)|^2 d^3 x d^3 y$.

If N is large and, as in our case, the number of excitations is sufficiently small compared to N, one may argue that the sums in the first and third term in (4.26) may be replaced by their respective mean-field values, which are constant due to the nature of Ω_t. This approximation is of course only justified for small coupling constants e^2, as starting from order e^4 also the state Ω_t will be perturbed due to the polarization terms. However, since our focus is on the ontology and not on finding the best approximation, we shall simply neglect this higher order polarization effect in the following. Hence, the remaining term (4.25) dictates the evolution of the electron-hole wave function η_t according to

$$i\partial_t \eta_t(x, y) = H_k^0(x_k)\eta_t(x_k, y_k) - \eta_t(x_k, y_k)\overline{H_k^0}(y_k) - U(x_k - y_k)\eta_t(x_k, y_k). \ (4.27)$$

Although a rigorous mathematical justification is lacking, one may have the well-founded hope that for the ansatz (4.19) with the approximated evolution of the vacuum state (4.16) and initial values (4.20), the right-hand side of (4.21) could in principle be estimated with Grönwall type estimate using our computation (4.24) to give a bound on the difference between Φ_t and $\Phi_t^{\approx}$—that is, the quality of our approximation for example in terms of a trace norm of the respective reduced density matrices.

In sum, we arrive at an economic two-particle description given by (4.20), (4.16) and (4.27) that approximates the actual N-particle dynamics of the initial excitation Φ in (4.17). As the initial vacuum state essentially does not change over time—cf. (4.16)—and since due to (4.9) the vacuum behaves as if there are no electrons, for all practical purposes, one may be inclined to forget about Ω entirely. The vacuum state Ω then simply encodes a state without excitations. A transition from Ω to Φ_t by means of (4.6), however, requires the introduction of the two-particle wave function η_t. Figure 4.2 illustrates these two views.

Further excitations require the introduction of further multi-particle excitations and so on. However, as long as the number of excitations is small compared to N, those excitation wave functions are much less complex objects to keep track off than the actual N-particle dynamics of Ψ_t.

To streamline the mathematics according to this idea, it makes sense to formulate the N-particle wave function dynamics (4.6) not on $\mathcal{H}^{\wedge N}$, the

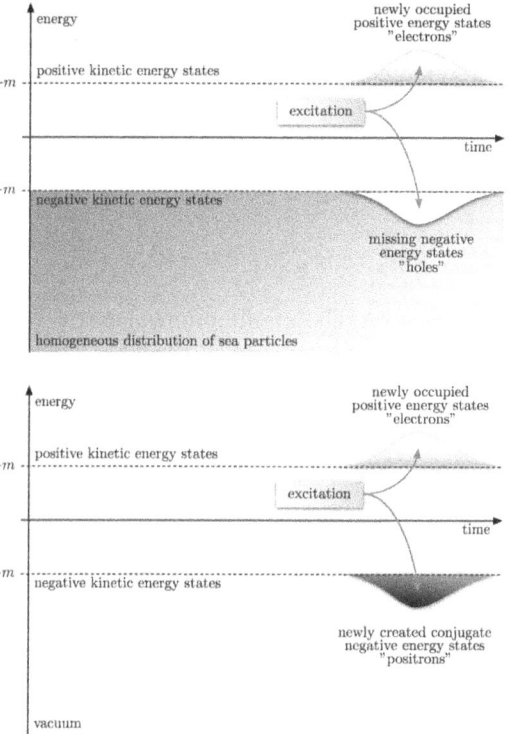

Figure 4.2 Both illustrations depict the same excitation of the wave function of the universe. It may have started out as the vacuum Ω with all negative energy states occupied, cf. (4.14), and then subject to the action of the external influence V, evolved into the excited universal wave function Φ given in (4.17). The top figure gives an account of this process in terms of the wave function of the universe. The approximate description of the evolution of the universal wave function Φ_t given by means of $\Phi_t^{\approx}$, cf. (4.19), is one in which the vacuum states Ω_t, cf. (4.16), and the excitation η_t, cf. (4.27), are evolved separately. As Ω_t is constant up to a phase, one may simply neglect it in the description and only account for the additional positive energy electron states and the states that have to be taken out of Ω_t in order to create the appropriate holes in the sea. This view of the universe is depicted in the bottom figure. As discussed in the text, this naive picture is only valid as long the vacuum polarization can be neglected—that is, the reaction of the created excited states back on the sea states that accumulates over time. Nevertheless, the view in the bottom figure can be maintained even when using for a reference state Ω_t, another one whose time evolution incorporates these polarization effects. This is also the view that is adopted in the standard formulation of QFT—see our comment (4.29) about the Ω-dependence of the representation.

Hilbert space of the N matter points that move electronwise, but on a space that only keeps track of the varying number of wave function excitations with respect to a reference state such as Ω much in the spirit of the bottom illustration in Figure 4.2. For this purpose one introduces the so-called Fock space $\mathcal{F}_\Omega$, which can be defined by employing the so-called creation and annihilation operators. Algebraically the creation operator ψ^* is given by

$$\psi^*(f)\,\varphi_1 \wedge \varphi_2 \wedge \dots \varphi_N = f \wedge \varphi_1 \wedge \varphi_2 \wedge \dots \varphi_N \qquad (4.28)$$

for any choice of $f, \varphi_k \in \mathcal{H}$. The annihilation operator ψ is defined as the corresponding adjoint ψ. Due to the anti-symmetric tensor product $\wedge$ these operators fulfill the anti-commutation relations

$$\{\psi(f), \psi^*(g)\} = \langle f, g \rangle, \qquad \{\psi(f), \psi(g)\} = 0 = \{\psi^*(f), \psi^*(g)\} \qquad \text{for all } f, g \in \mathcal{H},$$
$$\text{and, furthermore,} \quad \psi(\chi)\Omega = 0 = \psi^*(\varphi)\Omega \qquad \text{for all } \chi \in \mathcal{H}^+, \varphi \in \mathcal{H}^-. \qquad (4.29)$$

The Fock space $\mathcal{F}_\Omega$ can then be defined as the tensor product $\mathcal{F}_\Omega = \mathcal{F}_\Omega^e \otimes \mathcal{F}_\Omega^h$ of the electron Fock space $\mathcal{F}_\Omega^e$ and hole Fock space $\mathcal{F}_\Omega^h$, which are spanned by finite linear combinations of the form $\psi^*(f_1) \dots \psi^*(f_n)\Omega$ and $\psi(g_1) \dots \psi(g_n)\Omega$ for $f_k \in \mathcal{H}^+$ and $g_k \in \mathcal{H}^-$, respectively. In this algebraic construction the specific reference state Ω in $\mathcal{H}^{\wedge N}$ is hidden as Ω is encoded by $|\Omega\rangle = 1 \otimes 1$. It has to be emphasized that the independence of $|\Omega\rangle$ from Ω in this notation (as also suggested by the bottom illustration in Figure 4.2), however, is an illusion, since Ω is encoded in the relations (4.29)—acknowledging the fact that Ω is the Dirac sea with all states in $\mathcal{H}^-$ occupied. By construction, there is a well-known canonical isomorphism between $\mathcal{H}^{\wedge N}$ and the N-particle sector of $\mathcal{F}_\Omega^N$—that is, the subspace of $\mathcal{F}_\Omega$ spanned by states with equal numbers of electron and hole excitations. It reads

$$\iota : \mathcal{H}^N \to \mathcal{F}_\Omega^N, \quad \iota(f_1 \wedge f_2 \wedge \dots \wedge f_N) := \psi^*(f_1)\dots\psi^*(f_N)\psi(\varphi_1)\dots\psi(\varphi_N)|\Omega\rangle \quad (4.30)$$

for $f_k \in \mathcal{H}$. Thus, the approximate excitation $\Phi_t^{\approx}$ for (4.19) is represented in $\mathcal{F}_\Omega$ by

$$|\Phi_t\rangle = \int d^3x \int dy\, \eta_t(x, y)\psi^*(x)\psi(y)|\Omega\rangle. \qquad (4.31)$$

Using these translation rules implied by ι, we can recast the actual N-particle dynamics (4.6) generated by the Hamiltonian H^N in terms of creation and annihilation operators as

$$i\partial_t|\Psi_t\rangle = \tilde{H}|\Psi_t\rangle$$
$$\tilde{H} = \int d^3x\, \psi^*(x)\big(H_x^0 + V_t(x)\big)\psi(x) - \int d^3x \int d^3y\, \psi^*(x)\psi^*(y)U(x-y)\psi(y)\psi(x). \qquad (4.32)$$

This expression is the standard second-quantized Hamiltonian of quantum electrodynamics (QED) in Coulomb gauge when neglecting radiation. Note that in $\tilde{H}$ the creation and annihilation operators appear in pairs, which ensures that for an initial state in the N-particle sector of $\mathcal{F}_\Omega$ the generated dynamics also remains there.

In conclusion, for an initial wave function $\Psi_{t=0} \in \mathcal{H}^{\wedge N}$, there is a unique initial Fock vector $|\Psi_{t=0}\rangle := \iota(\Psi_{t=0})$ such that the time evolutions $(\Psi_t)_{t\in\mathbb{R}}$ on Hilbert space $\mathcal{H}^{\wedge N}$ and $(|\Psi_t\rangle)_{t\in\mathbb{R}}$ on Fock space $\mathcal{F}_\Omega$ generated by the Hamiltonians H^N in (4.6) and $\tilde{H}$, respectively, fulfill $|\Psi_t\rangle := \iota(\Psi_t)$ for all times t. In other words, the N-particle formalism is equivalent to the Fock space formalism when restricting to the N-particle sector. While the former represents the dynamics absolutely, the latter represents the dynamics with respect to a reference state Ω (see again Figure 4.2). The latter is, as we described above for the case of the two-particle excitation, tailor-made for finding approximations of excitations of Ω.

4.5 The appearance of particle creations and annihilations

In using the second-quantized formalism in section 4.4 instead of the N-particle formalism in (4.6), one is naturally led to a mode of language in which excitations are "created" and "annihilated". When an initial reference vacuum state Ω evolves into an excited state like Φ in (4.17), an excitation η_t is created, which effectively evolves according to (4.27). As can be seen from (4.27), the interaction between the electron and hole components of η_t is attractive, which may lead to a recombination unless an external influence $V(t, x)$ prevents it. In such a process the evolved excitation $\tilde{\Phi}_t$ decays back to the reference state Ω resulting in the annihilation of the excitation.

One may be inclined to refer to the electron component of η_t as describing one created "electron" and the hole component as describing one created "hole" or, synonymously, "positron". This, however, is an abuse of the well-defined term "electron" given by the introduced ontology—namely, a persistent particle that moves electronwise. Even if the state $\tilde{\Phi}_t$ in (4.19) were not only an approximation but the actual N-particle wave function, by the form of the velocity law (4.4) and anti-symmetry, the electron component of η_t would not influence only one electron, but all of them. It is, therefore, the collective motion of all electrons that causes the excitation η_t to appear. In order to develop a heuristic picture about this, let us go back to our initial analogy of the pressure waves in a classical gas. As we have discussed already, the pressure wave in the gas corresponds to the excitation η_t. The phenomenon of the pressure wave is perceived by the corresponding increase of density due to the configuration of gas molecules. In the same way, at a particular time t, it is the configuration of electrons in the Dirac sea that makes up the position statistics of the excitation η_t in the Dirac sea. Here, it is, however, important to note that the gas molecules do not necessarily move along with the dispersing pressure wave in the gas; instead, they oscillate. At different times, the phenomenon of the pressure wave is due to different molecules of the gas. Similarly, it is not necessarily the same electrons that make up the position statistics encoded in the excitation η_t. It would thus be a misconception to interpret η_t as guiding a two-particle system when the word "particle" is intended to refer to matter points existing in nature.

On the mathematical level of wave functions, the hole component of η_t simply encodes what has to be taken out of the reference state Ω_t, while the electron component of η_t encodes what has to be added to Ω_t in order to get a good approximation of the actual quantum state Ψ_t of the Dirac sea. This convenient approximation in terms of Ω comes at the price of losing direct information about the actual particle trajectories: as mentioned, in this case only the statistics about particle positions encoded in the wave function are under control. Hence, all that can be said about η_t is that against the uniform distribution of a sea of electrons described by Ω_t—due to the equilibrium condition (4.9)— $|\eta_t(x, y)|^2 d^3x\, d^3y$ is the probability of finding an additional negative charge and the absence of a negative charge (or mathematically equivalently, the presence of a positive charge) in the volumes d^3x at x and d^3y at y, respectively.

Nevertheless, the excitation η_t has the same properties as wave-packets and would in principle be capable of guiding two particles. To emphasize the difference between such excitations and actual particles, the term *quasi-particles* is usually employed in physics (as, e.g., phonons being the quasi-particles describing quantum excitations of crystal lattice oscillations). In this view, electron and hole components of η_t describe electron and hole quasi-particles having exactly the same properties except for the opposite charge, as can be seen from (4.27). The bottom line is that at all times there are N electrons in the Dirac sea. None of them were ever created or destroyed. It is their collective motion, which may deviate from the vacuum motion, that gives rise to what we refer to as "electron-positron pair-creation".

One can emphasize the quasi-particles character of the excitations already in the mathematical formalism. For this it is convenient to split the field operators ψ^*, ψ into electron and positron creation and annihilation operators

$$\psi^*(f) = b^*(f) + c^*(f) \quad \text{with} \quad b^*(f) = \psi^*(P^+f), \quad c^*(f) = \psi^*(P^-f)f \in \mathcal{H}, \quad (4.33)$$

where $P^\pm$ are the orthogonal projectors on $\mathcal{H}$ to $\mathcal{H}^\pm$, respectively. The number of electrons and hole quasi-particles can then be given by

$$N^e = \sum_n b^*(\chi_n)b(\chi_n), \qquad N^h = \sum_n c(\varphi_n)c^*(\varphi_n), \qquad (4.34)$$

using any orthonormal bases $(\chi_n)_{n=1,\ldots,N}$ in $\mathcal{H}^+$ and $(\varphi_n)_{n=1,\ldots,N}$ in $\mathcal{H}^-$.

Note that by virtue of (4.29), for instance,

$$N^e|\Omega\rangle = 0 = N^b|\Omega\rangle, \tag{4.35}$$

while for Φ in (4.17), $N^e|\Phi\rangle = 1 = N^b|\Phi\rangle$ as expected.

However, much caution has to be taken when interpreting the particle number operators in (4.34). First, these operators do not yield a definite number before taking, for instance, their expectation value. Even then, rather than an absolute number, these operators record only the number of wave-packet excitations—such as, for instance, η_t—relative to the chosen reference state Ω_t, as can be seen from (4.35)—that is, (4.29). Furthermore, recall that in the preceding section when considering the case of zero external influence—$V(t, x) = 0$—we argued that the ground state of the free theory, Ω, is a good candidate to approximate the vacuum. Let us stress again that this choice stemmed from the need to find a good approximation and was by no means unique (which is why we regarded a whole class of reference equilibrium states—recall that Ω was just a representative in the case of no external influence $V(t, x) = 0$). Depending on the kind of approximation in mind, one choice of reference state Ω might be better than another. Changing the reference state, however, naturally changes the meaning of the number operators, as excitations are only defined with respect to it.

For instance, when an external influence $V(t, x)$ is present, the equilibrium conditions—that is, conditions (a) and (b) discussed in (4.9) and later in this chapter—naturally change. When $V(t, x)$ only changes very slowly, physicists usually employ the so-called Furry picture: for each time t one chooses as reference Ω the state in which all negative energy states of the Hamiltonian $H^0(x) + V(t, x)$ are filled. However, it is shown in Fierz and Scharf (1979) that even this seemingly canonic construction of reference states (4.34) is not Lorentz invariant—in the sense that what might appear as an empty vacuum in one reference frame may contain many quasi-particles in another. This is because the spectrum, employed to define Ω, transforms like time under Lorentz boosts. Hence, it does not give rise to an invariant.

The choice of a reference state Ω will ultimately depend on physics in the sense that it concerns an experiment on the appropriate detector model that specifies what the equilibrium conditions (4.9) are, and hence, which excitations cause clicks. At temperature zero, a good candidate for a reference state fulfilling this condition surely is the interacting ground state corresponding to the given experimental setup. This is because the Fermi exclusion principle and the Coulomb repulsion will naturally act to distribute the electrons in a sufficiently uniform manner in order to meet (4.9). Mathematically, such states are very difficult to compute, as, due to the pair-interaction, these states are highly entangled instead of being product states. However, as is evident from the success of QED, in many situations these highly entangled ground states can for all practical purposes be approximated by product states such as Ω (e.g. in the sense of reduced density matrices as discussed earlier). It has to be emphasized that this need for a reference state is only an artifact of our approximation in order to deal with the complexity problem (II) of section 4.2. The Dirac sea model is independent of observers or detector models: there are always N electrons in the sea moving according to the laws (4.4) and (4.6).

Consequently, the choice of Ω should rather be seen as a choice of suitable coordinates with respect to which one may track the complicated N-particles dynamics (4.6) most conveniently. In principle, any choice is possible, since the isomorphism (4.30) between the N-particle formalism (4.6) and the formalism in terms of creation and annihilation operators (4.32) does not depend on particular properties of Ω. The number operators that are defined with respect to this choice Ω, however, can be assigned a physical interpretation only with respect to that particular Ω. For example, if Ω fulfills the equilibrium conditions (a) and (b) discussed in (4.9) and later in this chapter, the net sum of all pair potentials between the electrons vanishes and each electron effectively feels a constant potential and, thus, moves freely as if it were in a vacuum (a prime example of what could be called dark matter, as effectively there is no interaction except for maybe gravitation). The presence of an excitation of Ω disturbs those conditions, resulting in an effective interaction between all electrons in the sea with the electron-hole excitation η_t, which dictates the form of (4.27). The density $|\eta_t|^2$ then describes the distribution of an additional negative and positive charge with respect to the uniform background distribution of Ω. Only in this sense do the particle number operators record the number of in principle physically measurable excitations. This is what is gained from the introduction of an ontology of persisting particles: it is capable of explaining the import of all the relevant mathematical quantities of the theory.

The Unruh effect (Unruh (1976)) can also be understood in this way: the equilibrium conditions were chosen to find a good approximation of the complex N-particle dynamics and are at best Lorentz invariant. Switching to an accelerated frame will, in general, violate those conditions. Consequently, states Ω and Ω' describing a vacuum in a rest frame and an accelerated frame, respectively, may differ. Thus, Ω may appear in the accelerated frame as excitation of Ω' and a detector in the accelerated reference frame may therefore click. In sum, excitations cannot be given an absolute but only a relative meaning with respect to a certain equilibrium state chosen for our effective description. This is the reason why there is no point in according an ontological significance to these excitations in the form of particle number operators or quasi-particles—which are the ontological particles in the proposal of Bell (2004, ch. 19)—since there is no canonical choice for Ω.

Finally, let us briefly come back one last time to the removal of the cut-offs and the mathematical idealization of the thermodynamic limit of infinitely many particles. It is a fact that when $V(t, x)$ is, for instance, given by an electromagnetic four-potential with non-zero spatial components, in this limit the time evolution cannot be implemented on Fock space $\mathcal{F}_\Omega$ with respect to the vacuum Ω as defined in (4.14) (see Ruijsenaars (1977)). Even a gauge transformation for a gauge four-potential with non-zero spatial components cannot be implemented as a map on $\mathcal{F}_\Omega$ into itself. A way out of this dilemma is to construct Fock spaces $\mathcal{F}_{\Omega_{V(t)}}$ for all times t with respect to reference states $\Omega_{V(t)}$ that are appropriately chosen according to the external influence $V(t, x)$, instead of the fixed vacuum Ω. A detailed discussion of this fact goes beyond the scope of this book. In brief, the upshot is that the time evolution has to be implemented on time-varying Fock spaces in terms of a map $\mathcal{F}_{\Omega_{V(t_0)}} \to \mathcal{F}_{\Omega_{V(t_1)}}$ for initial and final times t_0, t_1 (see Deckert et al. (2010)). The restriction on the choice of admissible $\Omega_{V(t)}$ depends on $V(t, x)$; it is by no means unique. Hence, the particle number operators in the Fock spaces $\mathcal{F}_{\Omega_{V(t)}}$ refer to excitations of different Ω_t. This shows again that there is not one vacuum Ω and no absolute particle number operator, but only a "choice of coordinates" in terms of reference states $\Omega_{V(t)}$, which makes the mathematical discussion of the Dirac sea tractable. However, this poses no problem, as the physical meaning of what we call excitations is as clear as the physical meaning of the reference state, which is specified by a physical consideration such as the equilibrium condition discussed above—or more concretely, a detector model.

4.6 The merits of the Bohmian approach

Before concluding this chapter, two points are worth being addressed. In the first place, one may gain the impression that the introduced formalism based on a reference state Ω, which fills the spectral subspace $\mathcal{H}^-$ with actual particles that move electronwise and leave $\mathcal{H}^+$ empty, violates one of the much-celebrated symmetries of QED—namely, the so-called charge symmetry (a symmetry that is broken in the weak interaction). This is a misconception. Charge symmetry simply demands that the equations of motion remain invariant when the signs of the elementary charge e and the one of the external influence are inverted, which is the case (see (4.6) and (4.27) for the effective equation).

Furthermore, it has to emphasized that even in the non-interacting case, while the statistics about the particle positions (which are encoded in the wave functions of the excitations thanks to Born's law) are Lorentz invariant, the actual particle trajectories generated by the velocity law (4.4) are not. It is a well-known fact that the latter depend on a preferred foliation of Minkowski space-time—in the setup above, we have chosen equal time hypersurfaces. This circumstance is not an artifact of our model; it is generic to all relativistic versions of Bohmian mechanics. The reason is the manifestly non-local nature of quantum mechanics as demonstrated, for instance, by Bell's theorem and the subsequent experiments. Instead of adding this extra structure of a preferred foliation by hand, it is possible to introduce it in a Lorentz-covariant manner by taking it to be determined by the N-particle wave function itself, or equivalently the Fock space vector (see Dürr et al. (2013a)). We did not do so because this would have made the definition of the Dirac sea model and our arguments for a particle ontology unnecessarily opaque. Furthermore, the actual trajectories of individual electrons figure only in the ontology of our theory, while, due to problem (I) mentioned in section 4.2, they alone do not allow to infer predictions. It is rather the collective motion of all particles in the Dirac sea that generates the correct position statistics, which are encoded in the excitation wave functions by Born's law; the latter are accessible and Lorentz-invariant. Again, the lack of Lorentz invariance of the particle trajectories in Bohmian mechanics should not be counted against the theory, since these are necessary to obtain definite measurement outcomes, and no one has developed a dynamics of quantum systems that is both Lorentz invariant and that accounts for definite measurement outcomes (see also section 5.5).

Despite the still open desiderata, we have demonstrated that the predictions of the standard model in the electron sector with cut-offs and neglect of radiation arise naturally from a theory of N persistent particles. Creation and annihilation appear only in an effective description with respect to a chosen reference state. This programme is not only applicable to the electron, but to all fermion sectors.

In conclusion, let us stress again the parallelism with Bohmian mechanics for quantum mechanics: in both cases, the ontology is exactly the same—a fixed, finite number of permanent point particles that are characterized only by their positions—that is, the relative distances among them. Furthermore, the dynamics is of the same type—that is, a deterministic law describing the motion of the particles on continuous trajectories without any jumps. In both cases, the objective is to answer the questions of what there is in nature on the fundamental level (i.e. permanent matter points) and how what there is evolves (i.e. provide a dynamics for the individual processes in nature). From this ontology and dynamics then follow the formalisms of statistical predictions of both standard quantum mechanics and QFT. In this sense, these predictions are explained by this underlying ontology and dynamics.

The same goes for the solution to the measurement problem that hits QFT in the same way as quantum mechanics: matter points like the Bohmian point particles are admitted because they explain the measurement outcomes. For instance, the dots on the screen as recorded in the double slit experiment are made up by these point particles. However, this explanation is not achieved by the (primitive) ontology of particles alone, but by this ontology together with the dynamics (i.e. the guiding equation): it is the dynamics that provides for the stability of macroscopic objects, including pointer positions and dots on a screen. As Dickson (2000) convincingly argued, to obtain this stability, no properties of the particles over and above their position in space are needed, but a dynamics is that provides for trajectories such that there are stable macroscopic particle configurations.

Bearing these facts in mind, the solution to the measurement problem provided by Dirac sea Bohmian QFT is of the same type: here again, the Bohmian point particles are admitted because they explain the measurement outcomes. There always is a definite, finite number of particles moving on continuous trajectories, with some of these particles making up the macroscopic phenomena that we see. As discussed, we only have to be more careful to relate those particles and their dynamics to what we see with the naked eye such as the trajectories recorded in cloud chamber detectors. The latter are generated by deviations from the collective motion of the Bohmian particles in an equilibrium state such as the vacuum. Thus, to stress again, what explains these phenomena is not the mere fact of there being particle configurations, but the particle dynamics: it consists in this case of excitations from a sea of particles in an equilibrium state, with these excitations, as elaborated on earlier, implying a change of the motion of in principle all the particles in the sea. To sum up, what explains the traces in the cloud chamber are these excitations as affecting the motion of the particles in the sea.

In sum, the Bohmian approach works for QFT in the same way as for quantum mechanics, and the argument for this approach is the same in QFT and in quantum mechanics: a solution to the measurement problem that tells us what happens in nature on the level of individual processes and that spells out how what thus happens explains the macroscopic phenomena including measurement outcomes. The Bohmian approach vindicates the minimalist ontology of there being only matter points individuated by distance relations and the change of these relations also when it comes to our currently most advanced physical theory of matter—namely, the standard model of QFT.

5 Relationalism for relativistic physics

5.1 The challenge from relativistic physics

The minimalist ontology that we pursue in this book is based on the following two axioms:

(1) *There are distance relations that individuate simple objects—namely, matter points.*
(2) *The matter points are permanent, with the distances between them changing.*

The distance relations are conveniently represented in terms of a three-dimensional geometry such as Euclidean geometry, although neither the geometry being three-dimensional nor its being Euclidean are necessary. The change in the distance relations enables a representation in terms of an order that is unique and directed, with time being that order, although its metric depends on the choice of a subsystem of distance relations within the universal configuration of matter relative to which change is measured.

A physical theory introduces a dynamical structure in terms of dynamical laws (usually formulated as differential equations) in which various dynamical parameters figure that are attributed either to the matter points taken individually (e.g. mass) or to an entire configuration of them (e.g. an entangled wave function), in the last resort the whole configuration of matter points of the universe (e.g. the universal wave function). The task of the dynamical structure is to capture the change in the distance relations among the matter points in a manner that is both simple and informative about that change. Neither the geometry nor the dynamical structure belong to the ontology. The ontology is completely given by the two mentioned axioms.

This ontology rests on a crucial distinction between *variation* (axiom 1) and *change* (axiom 2). There is variation within a configuration in virtue of the distance relations being such that they discern—and thereby individuate—the matter points. And there is change in the distance relations, with that change providing for an intertemporal identity of the matter points, enabling a representation of that change in terms of the matter points moving on continuous trajectories. The argument for this ontology is its simplicity together with its empirical adequacy: through axiom 1, this ontology is most simple. If there is a plurality of objects, there has to be a relation of a certain type that unifies the world. If that relation also individuates the objects—so that no commitment to intrinsic properties, a primitive thisness or bare substrata is called for—the ontology is most simple. The distance relation satisfies this condition. It is at least not obvious what other type of concrete physical relation could do so as well. Through axiom 2, this ontology is empirically adequate. If there were only relations that individuate the fundamental objects, but no change, the ontology would be most simple. However, it would have no chance of being empirically adequate. The simplest way to introduce change is in terms of change of the relations that are there anyway to individuate the objects. This is sufficient to obtain empirical adequacy: all the empirical data consist in relative positions of objects and change of these positions (i.e. motion). That is why no ontological commitment neither to the geometry nor to the dynamical parameters that a physical theory employs in order to represent that change is necessary (which would require at least one additional axiom). The two mentioned axioms constitute a minimally sufficient ontology that is empirically adequate. In a nutshell, as argued in the preceding chapters, less won't do for an ontology of the natural world; bringing in more creates new drawbacks instead of providing additional explanatory value.

The geometry and dynamical structures considered in Chapter 3 (classical mechanics of gravitation, Bohmian quantum mechanics) and in Chapter 4 (Dirac sea Bohmian QFT) are such that the configuration of matter is represented as being inserted into a three-dimensional Euclidean space and as evolving in time such that the evolution of the configuration at a time t is fixed by dynamical parameters that are attributed to that configuration at that very t (e.g. particle masses and charges, or an entangled wave function). Nonetheless, for the geometry and the dynamical structure to achieve a description of the change in the configuration of matter that is most simple and informative, this representation has to be based on the change in the distance relations that actually occurs. In other words, the change comes first, and then come the geometry and the dynamical laws as means of representing that change. If the geometry and the dynamical structure are no addition to being, then the truth-maker for the propositions that ascribe geometrical as well as dynamical parameters such as particle masses and charges or a wave function to the configuration of matter points at a time t cannot be the distance relations at that very t. What makes these ascriptions true is the change that actually occurs in the distance relations among the matter points. The way in which that change occurs manifests certain patterns or regularities.

Because there are such patterns, a physical theory can achieve a description of that change that is most simple and informative by attaching geometrical and dynamical parameters to the configuration of matter points. It is a gain in simplicity if by attributing dynamical parameters to the configuration of matter points at a given time t and inserting these parameters as initial conditions into a dynamical law, one achieves a description of the whole past and future evolution of that configuration, although the truth-maker for this ascription is not located in the configuration of matter at that very t.

However, it is a contingent issue whether the patterns that the change in the distance relations among the matter points exhibits are such that this change can be represented in terms of a dynamical structure that requires only parameters attributed to the configuration of matter points at a given time as initial condition. As regards the ontology, the change in any one distance relation between two matter points entails a change of in principle all the other distance relations in the configuration. But this does not imply that when it comes to representing change, it is possible to obtain a representation that is both simple and informative about the change that actually occurs on the basis of initial conditions that consist in parameters attributed to the configuration of matter points at a given time. The correlated motions of quantum particles represented by means of an entangled wave function enable such a representation and are the reason why we have written down the fundamental dynamics of QFT in a non-relativistic manner in Chapter 4 (in the sense that this dynamics requires a notion of objective simultaneity).

Relativity physics challenges this manner of representing change. In relativistic interactions, it is dynamical parameters situated in the past (and possibly also in the future) that capture the evolution of a given configuration, whereby the distance relations in the given configuration are not relevant for the dynamical structure. That is why relativity physics not only introduces a new dynamical structure but also challenges the ontology as given by the two mentioned axioms, because this ontology is built on matter points that are individuated by the distance relations in a given configuration and the change in these relations. In this chapter, we address this challenge. We first consider classical electrodynamics, which is the first relativistic theory, arguing that fields are part of the dynamical structure of this theory instead of new elements of the ontology and discussing an alternative theory of classical electrodynamics that works with direct particle interactions (i.e. the Wheeler-Feynman theory) (section 2). We then argue in general terms that the minimalist ontology pursued in this book remains a cogent stance also when it comes to relativistic physics (section 3) and show how Super-Humeanism combined with this minimalist ontology is a valid option also for general relativistic physics (section 4). Finally, we briefly go into the relationship between quantum entanglement and relativity (section 5).

5.2   Fields and relativistic laws in classical electrodynamics

Since the advent of classical electrodynamics, the field concept has conquered modern physics, which today is to a large extent field theory. However, while fields have proven to be extremely successful as effective devices, physics and philosophy run into an impasse when they accord an ontological significance to these devices and buy into a dualistic ontology of fields and particles, or an ontology that replaces particles with fields. On closer examination, the concept of fields as mediators of particle interactions turns out to be philosophically unsatisfying and physically problematic, as it leads, in particular, to self-interaction singularities (see Lange (2002) for a good introduction to the debate about fields in physics and philosophy). Against this background, we argue that the true significance of fields is that of bookkeeping variables, summarizing the effects of diachronic relativistic interactions in order to obtain an efficient description of subsystems in terms of initial data (cf. Mundy (1989), p. 45). Indeed, objections that match the ones against absolute space in Chapter 2, section 1, apply to an ontological dualism of particles and fields as well: fields also stretch out to infinity far beyond where any particles are whose motion they could influence (see e.g. Feynman (1966), pp. 699–700). More importantly, the ontological status of fields is unclear. Are they properties—namely, properties of space-time points? Or are they some sort of stuff filling space-time?

If fields are properties of space-time points, one falls back to the commitment to an absolute space or space-time whose points instantiate the field properties (this commitment is clearly brought out by Field (1980), p. 35, and Field (1985), pp. 40–42). Over and above the drawbacks that the commitment to an absolute space or space-time implies, geometrical properties are *bona fide* properties of space-time, but it is rather odd to attribute in addition to the geometrical properties also causal properties influencing the motion of certain particles—namely, the charged ones—in the guise of field properties to the space-time points. To put it differently, if properties of space-time points are causal in the sense that they influence the motion of matter, then they should have an effect on the motion of *all* particles, not just the charged ones; after all, it is precisely the universality of gravitation that motivates its geometrical account in general relativity theory.

Moreover, if one conceives the field properties as dispositions, one faces the following problem: the manifestation of the original dispositional property of the particles—namely, their charge—which consists in the acceleration of other particles, then is mediated by further dispositional properties —namely, the field properties. Thus, a disposition (such as the charge of a particle) produces in the first place a further disposition (the field properties), and both these dispositions then manifest themselves in the acceleration of particles. Consequently, in any interaction mediated by a field, there are two distinct dispositional properties that bring about the same manifestation.

If one conceives fields as a stuff filling space-time, one faces the drawbacks of the gunk view of matter discussed at the end of Chapter 2, section 1—namely, the commitment to a bare substratum of matter that, moreover, admits different degrees at different points of space-time as a primitive matter of fact. In addition to that, in the case of the electromagnetic field, there can be points or regions of space-time where the field value is zero. Is there no field stuff in these regions? Or does the field stuff exist everywhere and merely exerts no force on particles in these regions?

Whereas the property view of fields usually goes with a dualistic ontology of particles and fields, the gunk view of fields is usually conceived as replacing particles with fields. However, as the discussion of the GRWm dynamics in Chapter 3, section 3, has shown, we do not have a convincing dynamics at our disposal that explains the experimental evidence (which is evidence of particle positions and motion) in terms of a field ontology, not to mention the atomistic constitution of matter from elementary particles via the chemical elements to molecules. This assessment holds independently of the issue of a dynamics in terms of wave function collapse for quantum mechanics: there simply is no clear-cut dynamics for a field ontology to explain the experimental evidence; the GRWm dynamics is the most advanced proposal in that respect. In sum, again, enriching the ontology—with fields in this case—leads to new drawbacks instead of providing additional explanatory value.

Consequently, the electromagnetic field has a dynamical role, but not an ontological status. That is to say: as we developed the Super-Humean strategy for geometrical space, mass, charge, etc., in classical mechanics and the wave function in quantum mechanics, so this strategy applies to fields as they appear in classical electrodynamics. They are part of the dynamical structure of the theory, being a means to describe the change in the particle positions (i.e. the change in their relative distances) in a manner that achieves the best combination of being simple and being informative about that change.

The proposal of a pure particle ontology runs counter to the intuition that the existence of at least certain forms of electromagnetic fields is somehow obvious. After all, electromagnetic fields are obviously there when we turn on the radio. Moreover, there seems to be a quite literal sense in which *all* we actually *see* is light. And light can obviously be manipulated: it can be reflected, refracted, polarized, absorbed . . . Against this background, our claim that light does not exist, since the electromagnetic field does not exist, may appear absurd. In fact, what we propose is to quine light, to borrow an expression from Dennett (1988) coined for the philosophy of mind ("Quining qualia").

Our claim is that propositions using the concepts of *light, radiation, electromagnetic signals*, etc., are perfectly true, but their truth-maker are particle motions only. This, again, is the Super-Humean strategy spelled out in Chapter 2, section 3. The "electromagnetic signals" that we pick up when we turn on the radio thus refer to a particular kind of interaction between our receiver and coherently oscillating particles at the broadcasting station. Our impression of "red light" refers to a particular kind of interaction between the observed object and our visual receptors. The "reflection of light" refers to a series of interactions involving at least a source S, a mirror M and a target T so that certain counterfactuals about the strength of the effect on T depending on the presence/absence of M and the geometry of the setup are true, etc. It is somewhat tedious, but quite straightforward to spell out the details.

In any case, science pushes us in the direction of this eliminative path. Our physical theories do not contain any concepts of *redness* or *blueness* or *greenness*. To the contrary, sense impressions involving the perception of colour are related to certain wave-lengths in the electromagnetic field. To eliminate the electromagnetic field as well in favour of an ontology of direct particle interactions then merely means to cut out the middle-man: the frequencies that we usually identify with red light or blue light or green light are taken to refer directly to accelerations of particles. Adopting the terminology of Sellars (1962), we can say the following: the *manifest image* and the *scientific image* of the world connect at other points than those ones proposed in the field theory, but this does not make the manifest image less accurate or the theoretical account less compelling.

Note that we are not concerned here with the phenomenology of light and colour sensation (that is, we are not concerned with "quining qualia"). Our point only is that the field ontology and the sparse particle ontology of the natural world are in the same boat when it comes to the link between the scientific image of the world and our sense impressions. Field ontology is not in a better position to establish that link than a sparse particle ontology, but, to the contrary, faces the mentioned drawbacks. Generally speaking, enriching the ontology of the natural world with entities that are not needed for a sparse and empirically adequate ontology does not help when it comes to the link between the natural world and the phenomenology of mental states, but only entails problems such as the mentioned ones in the case of an ontological commitment to fields.

One can also formulate gravitation in Newtonian mechanics in terms of a gravitational field. But in this case, it is obvious that the gravitational field (or potential) is just a useful mathematical tool to represent the direct particle interactions rather than a candidate for ontology. The reason is that the degrees of freedom contained in the gravitational field at any time t can be immediately reduced to the particle configuration at the same time. The electromagnetic field at time t, by contrast, depends on the trajectories of the particles in the past (and possibly also the future) of t, because the sources affect the field in a retarded (and possibly also an advanced) way, reflecting the light cone structure of relativistic space-time. This fact is no reason to elevate the electromagnetic field to a different ontological status than the gravitational field in Newtonian mechanics. But it explains why the field is harder to dispense with in the relativistic case: it is a book-keeper of past events—and possibly also future events—rather than a summary of co-present events.

The retardation furthermore explains the *appearance* that electromagnetic effects propagate at finite speed as if they had *sit venia verbo* a life of their own. After all, when we look up to the night-sky, we see many stars that have long ceased to exist. So, one might say, whatever interacts with us to create the impression of the star is not the star itself (which exploded millions of years ago) but something else—namely, electromagnetic radiation, which had been emitted from the star and then traveled all the way to earth. However, while it is a natural way of speaking to say that electromagnetic effects "travel" or "propagate", the ontological conclusion that there must thus be *something*, in addition to the particles, that actually propagates through time and space is unwarranted. The regularities in a world may be such that the present motion of subsystems of the universe is correlated with the present motion of distant subsystems (as in Newtonian mechanics), or that it is correlated with the past (and possibly future) motion of other subsystems intersecting the past (and possibly future) light cones of the subsystems (as in relativistic theories). This depends on what the laws turn out to be that provide for the best system, achieving the best balance between being simple and being informative about the actual motion conducted by the systems in the universe. But all this supervenes on the Humean mosaic of particle distances and their change only.

Another important difference between the electromagnetic field and the gravitational field in Newtonian mechanics is that the Maxwell equations allow for a variety of non-trivial and non-equivalent vacuum solutions—that is, solutions that do not have charged particles as field sources. These so-called *free fields* thus cannot be expressed in terms of the charge-trajectories; they do however affect the electromagnetic forces and hence the particle motions. The existence of free fields is sometimes voiced as an objection against the viability of a pure particle ontology for classical electrodynamics. However, to emphasize again, the Super-Humean strategy is not committed to mathematically reducing all dynamical structures to the primitive variables. Rather, dynamical structures such as free fields can enter as part of the best system description of the Humean mosaic, while the latter consists only of particle distances and their change.

That said, the existence of free fields can hardly be maintained on empirical grounds. Free fields are certainly necessary for providing an efficient description of subsystems. When we set out to describe the dynamics of N charges in a space-time region M, we have to account for the influence of charges outside of M by specifying appropriate boundary conditions for the electromagnetic field (the "incoming radiation"). These boundary conditions will in general contain free fields, or rather external fields, in the sense of fields that have no sources in M (because they have sources outside of M). However, on the fundamental level—that is, when M is the whole of space-time—the assumption of genuinely free fields coming in "from infinity" seems unwarranted. Empirically, we will never be able to determine that some observed radiation is truly source-free. In fact, good scientific practice is to assume that it is *not* and look for—or simply infer—the existence of material sources. Empirical evidence is fully consistent with the assumption that the model of classical electrodynamics that best fits our universe is one in which all field degrees of freedom can be ultimately reduced to the history of charged matter. However, the models of classical electrodynamics (without self-interaction) in which all field degrees of freedom can be reduced to the history of matter are also models of a corresponding direct interaction theory. On the fundamental level, it is thus unnecessary and unwarranted to buy into free fields as a physical possibility.

Whereas the philosophical problems that an ontological commitment to fields entails can be removed by considering fields as parts of the dynamical structure instead of the ontology of classical electrodynamics, there are physical problems in the Maxwell-Lorentz theory that cannot be solved in that way, requiring in fact new physics. The situation hence is different from the one in classical mechanics: Leibnizian relationalism can be vindicated as an ontology for Newtonian mechanics, which is a perfectly consistent physical theory, by adopting Super-Humeanism with respect to the geometrical and dynamical structure of Newtonian mechanics. One can also develop a relationalist alternative physical theory (as Barbour does), but this is by no means mandatory; in any case, also an alternative physical theory requires more conceptual means than the ones provided by the sparse ontology of Leibnizian relationalism. In the Maxwell-Lorentz theory of electromagnetism, to the contrary, there is a physical problem—namely the one of self-interactions—which constitutes a physical motivation to search for an alternative physical theory, independently of the problems that come with an ontological commitment to fields.

The Maxwell-Lorentz theory of classical electrodynamics—as the name already suggests—consists of two parts. The Maxwell equations describe the evolution of the electromagnetic field and the coupling of the field to charges and currents. The Lorentz force equation describes the motion of a charged particle in the presence of an electromagnetic field:

$$m\ddot{x}^{\mu} = e\, F^{\mu\nu}\, \dot{x}_{\nu}. \tag{5.1}$$

Here, $F^{\mu\nu}$ is the field tensor, comprising electric and magnetic components, m denotes the mass and e the charge of the particle with space-time trajectory $x^{\mu}(\tau)$ and a dot denotes a derivative with respect to eigentime τ.

The Maxwell field equations can be separated again into homogeneous and inhomogeneous equations, where the first involve only field degrees of freedom. The homogenous equations tell us that the anti-symmetric field tensor $F^{\mu\nu}$ (a two-form) can be written as the exterior derivative of a potential (a one-form) A^μ—that is, as

$$F^{\mu\nu} = \partial^\mu A^\nu - \partial^\nu A^\mu. \tag{5.2}$$

The inhomogeneous Maxwell equations couple the field degrees of freedom to matter—that is, they tell us how charges influence the electromagnetic field. Fixing the gauge-freedom contained in (5.2) by demanding $\partial_\mu A^\mu(x)$ = 0 (Lorentz gauge), the remaining Maxwell equations take the particularly simple form:

$$\Box A^\mu = 4\pi j^\mu, \tag{5.3}$$

with $\Box = \partial_\mu \partial^\mu$ the d'Alembert operator and j^μ the four-current density, which for N point charges is:

$$j^\mu(x) = \sum_{i=1}^{N} j_i^\mu = \sum_{i=1}^{N} e_i \int \delta^4(x - z_i(\tau_i)) \dot{z}_i^\mu(\tau_i) \, d\tau_i. \tag{5.4}$$

Now, *given* the trajectories $z_i(\tau_i) i = 1, \ldots, N$ of the particles, the solutions of (5.3) are well known. By the linearity of equation (5.3), we can sum the contribution of each particle. A special solution is given by the advanced and retarded *Liénard-Wiechert* potentials:

$$A_{i,\pm}^\mu(x) = e_i \frac{\dot{z}_i^\mu(\tau_i^\pm)}{(x^\nu - z_i^\nu(\tau_i^\pm))\dot{z}_{i,\nu}(\tau_i^\pm)}, \tag{5.5}$$

where x denotes a space-time point and $\tau_i^+(x)$ and $\tau_i^-(x)$ are the *advanced* and *retarded* times given as implicit solutions of

$$(x^\mu - z_i^\mu(\tau))(x_\mu - z_{i,\mu}(\tau)) = 0. \tag{5.6}$$

This means that the field equation connects events with zero Minkowski distance. In other words, it means that the advanced / retarded Liénard-Wiechert field at x depends on the charge-trajectories at their points of intersection with the future, respectively the past light cone of x.

To any such solution, we can further add any solution of the free wave equation

$$\Box A^\mu = 0, \tag{5.7}$$

corresponding to the aforementioned free fields.

The problem with classical electrodynamics is that a self-consistent description of an N-particle system requires us to solve (5.3) and (5.1) *together*. But this set of coupled differential equations is ill-defined. The Liénard-Wiechert field (5.5) is singular precisely at the points where it has to be evaluated in (5.1)—namely, on the worldlines of the particles. This is the notorious problem of the *electron self-interaction*: a charged particle generates a field, the field acts back on the particle, and since the field-strength is infinite at the position of the particle, the interaction blows up. As mentioned at the beginning of Chapter 4, this problem carries on to QFT in the guise of the ultraviolet divergences. Note that already the 1-body problem is ill-defined in the Maxwell-Lorentz theory. It is not the interaction between particles, but the duality of particle and field that leads to singularities. The reason why classical electrodynamics still works so well for most practical purposes is that physicists, in general, solve either the Maxwell equations for a given charge-distribution or the Lorentz equation for a given electromagnetic field, but not both together. Nonetheless, strictly speaking, the Maxwell-Lorentz field theory is an inconsistent theory.

One prominent way to cope with the self-interaction problem is implemented in the Lorentz-Dirac theory. The equation that is believed to describe all radiative phenomena correctly is the Lorentz-Dirac equation

$$m\ddot{x}^{\mu} = eF^{\mu\nu}\dot{x}_{\nu} + \frac{2}{3}e^2(\dot{x}^{\mu}\dddot{x}^{\nu}\ddot{x}_{\nu} - \dddot{x}^{\mu}), \qquad (5.8)$$

where $F^{\mu\nu}$ does not contain the self-field. The four-vector

$$\Gamma^{\mu} := \frac{2}{3}e^2(\dot{x}^{\mu}\ddot{x}^{\nu}\ddot{x}_{\nu} - \dddot{x}^{\mu}) \qquad (5.9)$$

is known as the *Schott-term* and includes, in particular, the radiation friction. Note that this equation of motion depends on the *third* derivative of $x(\tau)$—that is, the time-derivative of the acceleration.

In his seminal paper on the classical electron theory, Dirac showed that (5.8) can be derived from Maxwell's equations together with the principle of energy-momentum conservation; analogous results can also be obtained by considering the point-particle limit for a spherical charge-distribution (Dirac (1938); see Rohrlich (1997) for a good historical overview). These derivations, however, have to rely on a highly dubious *mass renormalization* procedure: an infinite, negative bare mass for the point-particle has to be assumed in order to cancel a diverging inertial term arising from the infinite self-energy. Moreover, while these derivations suggest that the self-interaction is the origin of the radiation reaction, this is not a consistent interpretation of the Lorentz-Dirac theory. The right-hand side of (5.8) is divergence-free at the position of the particle and hence does not contain any self-field according to Maxwell's equations (cf. Wheeler and Feynman (1945)). Nevertheless, in the end, we do not have to take the derivation seriously in order to accept the result. If equation (5.8) proves to be meaningful and empirically adequate, we may accept it as the correct law of motion for the classical electron regardless of its logical relationship to the original Maxwell-Lorentz theory.

Unfortunately, also the Lorentz-Dirac theory faces severe issues: (i) except for very fine-tuned initial conditions, the infamous triple-dot term in equation (5.8) leads to *runaway solutions* that diverge in finite time due to particles (self-)accelerating to the speed of light (see Spohn (2000) and the discussion in Frisch (2005)). In this sense, the infinite self-energy—respectively, the negative bare mass introduced to cancel it—still manifests as the source of dynamical instabilities. (ii) It is essential that the self-field of the particle does not appear on the right-hand side of (5.8). Consequently, the Lorentz-Dirac theory involves not just *one* electromagnetic field, but a different type of electromagnetic field for each particle in the universe. The field created by particle j must carry some property that makes it interact with all the other charges, but not with j itself. This leads to an enormous increase in mathematical complexity of the N-particle problem (we have to specify N initial fields and track the evolution of each one separately) as well as to a grotesque inflation of the physical ontology.

This situation motivates the search for a formulation of the dynamical structure of classical electrodynamics without the electromagnetic field. This idea goes back to Gauss (1867) and was further explored by Schwarzschild (1903), Tetrode (1922) and Fokker (1929). Wheeler and Feynman (1945) showed in their absorber theory that a time-symmetric direct interaction formulation of classical electrodynamics is able to account for all radiative phenomena. Henceforth, the respective direct interaction theory is commonly known as *Wheeler-Feynman electrodynamics*. This theory involves no self-interactions from the beginning. The Lorentz-Dirac equation including, in particular, the radiation friction can be derived from a statistical analysis as a phenomenological description. Moreover, there are good indications that the Wheeler-Feynman theory is free of runaway solutions (Bauer (1997)), which is to say that the solutions of the Lorentz-Dirac theory that are also approximate solutions of Wheeler-Feynman are automatically the good ones that do not lead to runaway singularities. In a nutshell, the Wheeler-Feynman theory captures precisely the physical content of the field theory while avoiding the unphysical artifacts.

A retarded direct interaction theory, such as the one proposed by Ritz (1908), cannot explain the phenomenon of radiation damping—that is, the fact that accelerated charges lose energy-momentum (unless one admits an *ad hoc* modification of the equations of motion like Mundy (1989)). In the absence of self-fields, this radiative reaction can only come from interactions with other charges, so that the damping effect would be considerably delayed in a purely retarded theory. This leaves us with the time-symmetric direct interaction theory proposed by Fokker (1929) and Wheeler and Feynman (1945). This theory is in a better position to explain the radiation damping, because advanced reactions to retarded actions arrive instantaneously.

Another important virtue of the time-symmetric theory is that it can be defined by a principle of least action for what is arguably the simplest relativistic action for interacting particles:

$$S = \sum_i \left[-m_i \int \sqrt{\dot{z}_i^\mu \dot{z}_{i,\mu}} \, d\lambda_i - \frac{1}{2} \sum_{i \neq j} e_i e_j \int \int \delta\big((z_i - z_j)^2\big) \dot{z}_i^\mu \dot{z}_{j,\mu} \, d\lambda_i \, d\lambda_j \right]. \quad (5.10)$$

The equations of motion for an N particle system then read as follows:

$$m_k \ddot{x}_k^\mu = \sum_{j \neq k} e_k e_j \frac{1}{2} \big({}^{(j)}F_{ret}^{\mu\nu} + {}^{(j)}F_{adv}^{\mu\nu} \big) \dot{x}_{k,\nu}, \quad (5.11)$$

where $F_{ret}^{\mu\nu}$ and $F_{adv}^{\mu\nu}$ correspond to the retarded and advanced Liénard-Wiechert fields, respectively. These equations contain no self-interaction, hence no singularities, and no mysterious free fields. The forces acting on a particle k are completely determined by the trajectories of other particles. This Wheeler-Feynman theory of classical electrodynamics faces notably two challenges—namely, to account for (i) the phenomenon of radiation damping and (ii) for the radiative arrow of time—that is, the fact that we observe only retarded radiation.

Wheeler and Feynman (1945) present a statistical account of the radiative reaction. They assume that an accelerated charge interacts with a large, homogeneous, spherically symmetric charge-distribution in the future. In the absence of external disturbances, the net force exerted by this so-called *absorber*, surrounding our local, low-entropy environment, is assumed to be approximately zero. Then Wheeler and Feynman show in a series of three computations of increasing generality that, if the absorber particles are disturbed by the retarded forces from the accelerated charge, their *advanced* reaction will correspond, in form and magnitude, to

$$\frac{1}{2}(F_{ret} - F_{adv}). \quad (5.12)$$

A test-particle in the vicinity of the accelerated charge will thus experience a net effect of

$$\frac{1}{2}(F_{ret} + F_{adv}) + \frac{1}{2}(F_{ret} - F_{adv}) = F_{ret}, \quad (5.13)$$

as if the accelerated charge produced a purely retarded force. Moreover, at the location of the charge, the difference $\frac{1}{2}(F_{ret} - F_{adv})$ corresponds precisely to the radiation reaction term (5.9), as was already shown by Dirac. In particular, the accelerated particle will thus experience a damping force as a result of its interaction with the absorber.

Apart from the justification of the temporal asymmetry thus introduced, the Wheeler-Feynman analysis of radiation reaction could have—and perhaps should have—ended here. The fact that it did not exposed their work to a lot of unnecessary criticism. The absorber response (5.12) is, notably, independent of any detailed properties of the absorber like mass or charge or the exact arrangement of its constituting particles. For this reason, Wheeler and Feynman—having already derived the absorber response in a series of hands-on computations—suggest that it should be possible to conclude the same result from first principles. Hence, they go on to present a remarkably simple and elegant argument based on the so-called absorber condition:

$$\sum_k \left(F_{ret}^{(k)} - F_{adv}^{(k)} \right) = 0, \tag{5.14}$$

where the sum goes over all particles in the universe. This amounts to the assumption that the distribution of charges in the universe form a complete absorber, so that $\sum_k F_{ret}^{(k)}(x) = 0$ and $\sum_k F_{adv}^{(k)}(x) = 0$ hold separately, at any space-time point x.[k]

One can imagine universes in which this assumption is valid to a good approximation, such as a compact spatial geometry with a homogeneous distribution of charges. But it is easy to point the finger at equation (5.14) and ask: "Why should we believe in that?". Many commentators have thus rejected the Wheeler-Feynman account, suggesting that the derivation of the radiative reaction rests on the validity of equation (5.14) (see, e.g., Zeh (2010), pp. 36–37, and Earman (2011), p. 368). This suggestion, however, is dubious, because—as shown by Wheeler's and Feynman's (1945) series of three computations and as further emphasized in Bauer et al. (2014)—the radiative reaction follows from statistical assumptions alone that are much weaker and much more robust than the infamous absorber condition.

While the calculations of Wheeler and Feynman seem sound and reasonable, one worry that immediately arises is that the same arguments apply also in the opposite time direction. If we assumed that the accelerated charge interacts (by advanced forces) with an absorber in the distant past, the retarded absorber response would correspond to $\frac{1}{2}(F_{adv} - F_{ret})$, resulting in a net force of F_{adv} on surrounding particles and an *anti-damping* force on the accelerated charge.

Since radiation damping is a many-particle phenomenon in Wheeler-Feynman electrodynamics, it makes sense to look for a thermodynamic explanation of the asymmetry. Unfortunately, for reasons explained next, the state space of the theory is not yet understood well enough to spell out such an account in detail. The general argument proposed by

Wheeler and Feynman (1945), which seeks to rule out a retarded response from the past absorber on probabilistic grounds, is generally not considered to be successful (see Arntzenius (1994), pp. 40–41, Price (1996), pp. 65–73, and Frisch (2005), ch. 6, for a critique, as well as Bauer et al. (2014) for a more positive assessment). However, while we do not yet have a conclusive account of the radiative asymmetry in Wheeler-Feynman electrodynamics, there are clear indications how the retrocausal effects manifested on the microscopic level can be reconciled with our macroscopic experience and the perceived arrow of radiation. This puts the Wheeler-Feynman theory in a position that is not worse than that of the field theory, where the explanation of the radiative asymmetry is subject to debate as well (for the current state of that debate see, e.g., Earman (2011)). Moreover, even if the justification of the asymmetry is an open question, the statistical derivation of the radiation reaction is already a significant success that the field theory cannot match, since there, one has to rely on a highly unphysical mass renormalization procedure in order to obtain the analogous result.

Arguably, the real reason why the Wheeler-Feynman theory is not appreciated by working physicists is this one: its equation of motion is not the kind of ordinary differential equation that physicists and mathematicians are trained to solve, but a so-called delay differential equation. As we can see from the action functional (5.10) (or, alternatively, from (5.11) together with (5.5)), the force acting on a particle at some space-time point x depends on the trajectory of the other particles at their points of intersection with the past and future light cones originating in x. The force is hence not determined by an instantaneous state of the system, where "instantaneous state" means the configuration of the physical system on a suitable space-like hypersurface that includes x.

As a result, the Wheeler-Feynman laws of motion cannot be posed as initial value problems. As of today, it is not known what kind of initial or boundary conditions one has to impose in order to ensure existence and uniqueness of solutions. The best conjecture is that initial data for a Wheeler-Feynman system comprise entire *segments of trajectories*, rather than instantaneous Cauchy data on a space-like hypersurface (see Bauer et al. (2013), Deckert et al. (2014), Deckert and Hinrichs (2016) and Bauer et al. (2016) for the current state of the solution theory). From this point of view, it is precisely the role of electromagnetic fields to introduce additional degrees of freedom that enable the formulation of relativistic laws as initial value problems. Indeed, there is no cogent reason to expect Cauchy data in a relativistic theory. Relativistic space-time, in contrast to Newtonian space-time, does not come equipped with a foliation into instantaneous spatial geometries. The dynamical relations in relativistic theories are typically spatio-temporal ones. Hence, we have no grounds to expect that the dynamical state of the universe can be completely described by physical data on a space-like hypersurface.

The field theory allows us to trade a diachronic, spatio-temporal description in terms of particle histories for a synchronic description in terms of an infinite number of field degrees of freedom that encode the history of charge-trajectories in their spatial dependencies. One can therefore say that the central significance of the fields, their *raison d'être*, is to serve as bookkeepers for the particle histories in order to save the concept of instantaneous dynamical states and the Newtonian paradigm of laws as initial value problems in a relativistic setting. Ultimately, though, this remains unsuccessful, since the initial value problem is not well-posed in the Maxwell-Lorentz theory either, even if self-interactions are neglected (see Deckert and Hartenstein (2016)). The reason is, simply put, that the initial fields specified on a spacelike hypersurface would have to include, in particular, the fields created by the charges in the past. However, we do not know what these fields are, unless we compute them from the charge trajectories in the first place. As Deckert and Hartenstein (2016) show, a generic choice of initial data—compatible with the Maxwell constraints—will lead to singularities in the field solutions which, in turn, produce singularities or discontinuities in the charge trajectories. Hence, even in the case of the field theory, the best way to obtain *physical* solutions of the fundamental laws is to abandon the initial value formulation and solve a delay differential equation in terms of the particle trajectories, of which the Wheeler-Feynman law (5.11) is a special case.

For practical purposes, the Wheeler-Feynman theory also allows us to introduce fields as bookkeeping variables in the sense discussed earlier. For a given distribution of charges, the effective description of a subsystem then corresponds to a boundary value problem within the usual Maxwell-Lorentz theory (respectively the Lorentz-Dirac theory, if the radiation friction is taken into account). The Wheeler-Feynman theory can therefore explain the success of textbook electrodynamics. On a fundamental level, the theory reminds us however that in relativistic physics, we have to expect the dynamics to depend on past and future particle motion rather than an instantaneous dynamical state of the universe.

Consequently, since the Wheeler-Feynman laws of motion draw on the geometric structure of relativistic space-time, with particles interacting along light cones, this raises the question how such a theory can be compatible with an ontology of distance relations that denies the fundamentality of both relativistic space-time and spatio-temporal relations. If relativistic laws of the Wheeler-Feynman type arise as part of the Humean best system, this means in the first place that the summary of the Humean mosaic that strikes the optimal balance between simplicity and strength is one in which the change of the particle configuration at one time depends on the particle configuration at former and later times (with "time" here referring only to the order of change in the configurations). This dependence is, however, only descriptive and does not entail or presuppose any real physical connection between particle configurations at different times.

Furthermore, the fact that relativistic laws of the Wheeler-Feynman type connect precise particle events with zero Minkowski distance means that there exists an embedding of the distance relations into a three-dimensional space, respectively a four-dimensional space-time, and a parameterization of the change of the distance relations (providing a temporal metric) such that the best system description relates precisely those particle events— that is, those vertices of the embedded network of relations for which the ratio of spatial distance and temporal distance is constant (that constant being the speed of light). It is plausible that a geometric representation for which this ratio is constant for dynamically related events would turn out to be more simple (and at least equally informative) as a representation in which it is not.

Hence, there is an embedding of the history of distance relations into a four-dimensional space-time geometry such that the best system laws connect those and only those particle events that are, in the usual nomenclature, light-like separated. Such laws will then naturally turn out to be invariant under symmetries that leave light-like connections—that is, the light cones—invariant. These symmetries are, of course, exactly those described by the Lorentz group. The space-time substantivalist will usually claim to provide an explanation of such symmetries in terms of the underlying space-time manifold. However, as mentioned already in Chapter 2, section 1, this does not amount to any deeper explanation, because the substantivalist, in turn, has to accept other structures and entities such as space-time points and the Minkowski metric as primitive.

In any case, one important consequence of the Lorentz symmetries is that, given a complete solution of the laws, there are an infinite number of equivalent possibilities—corresponding to different Lorentz frames—to slice up the space-time trajectories into a history of spatial relations. Any such foliation of four-dimensional space-time into three-dimensional geometries will identify different instantaneous configurations with different spatial distances between the matter points. However, basing oneself on an ontology of distance relations, at most one of these foliations can accurately represent the real state of affairs, that is, identify the correct co-present configurations and a family of three metrics such that the spatial distances between matter points correspond to the fundamental distance relation. The best system laws do not identify the "correct" frame of reference.

5.3   Relativistic dynamical structure and minimalist ontology

When it comes to the relationship between on the one hand ontology and on the other hand geometry and dynamical structure, the situation in relativistic physics is the same as in Newtonian mechanics: simplicity in ontology and simplicity in representation pull in opposite directions. If one starts from the dynamical structure of Newtonian mechanics, being the paradigm example of a simple and universal theory, and uses the dynamical structure as guideline to the ontology, one gets to an ontology of point particles, possessing mass and moving in an absolute space and an absolute time, because inertial motion in Newtonian mechanics is defined as motion with respect to an absolute space in an absolute time. However, as Leibniz, Mach and others point out, this ontology faces serious drawbacks, notably the three ones discussed in Chapter 2, section 1 (differences that do not make a difference, infinite expansion and problem of what characterizes matter as filling space). Given that all the empirical evidence consists in relative particle positions and change of these positions, the relationalist ontology as stated in the two axioms on which this book is based is the more simple ontology that is empirically adequate and that avoids these drawbacks, which originate in the surplus structure that comes with endorsing an absolute space and time.

In relativistic physics, the situation is similar. Starting from a relativistic dynamical structure as given by the Maxwell-Lorentz theory of classical electrodynamics or the Wheeler-Feynman theory and implemented in the geometry of the special theory of relativity, there are now not only two, but three main candidates for an ontology that clarifies the relationship between matter and space-time.

(i) In the first place, there is the option corresponding to the Newtonian one. There are point particles moving in an absolute space, with each particle being individuated by its position in absolute space. Consequently, distance relations to other particles in a given configuration are not relevant for the individuation of the particles. These relations derive from the geometry of the underlying space. The particles are unified by being situated in one absolute space. That space is constituted by the topological, affine and metrical connections among its points. That ontology hence is the same as in Newtonian mechanics. The availability of this ontology highlights again the importance to distinguish between what individuates the particles and what concerns the dynamical structure that captures their motion. On this ontology, the individuation is carried out by the particle position in absolute space so that simultaneity plays no role for the individuation of the particles. Hence, absolute space has a crucial function for the particle ontology, whereas the nature of time is an issue of the dynamical structure.

As regards the dynamical structure, in the relativistic case by contrast to the Newtonian one, the change of the position of a given particle is not fixed by the simultaneous positions and velocities of the other particles in a given configuration (e.g. via a law of gravitation based on attributing a mass and an initial velocity to the particles). In a field theoretical setting, the change of the position of a given particle is determined by the field configuration in its immediate past light cone (via a field law based on attributing a charge to the particle in addition to its initial velocity and inertial mass, with that charge reacting to the field configuration in its immediate past, and that field configuration, in turn, being fixed in the last resort by other charged particles in the past light cone—leaving aside free fields, as argued in the preceding section). In a setting of direct particle interaction as in the Wheeler-Feynman theory, the change of the position of a given particle is determined by the past and future motions of other particles— namely, the motions of the other particles in its past and future light cones (via a law of electrodynamics based on attributing a charge over and above an inertial mass and an initial velocity to the particles, but no field configuration). In this latter case, there is empty space between the particles as in Newtonian mechanics, whereas in the former case, space is represented as being filled with fields. Consequently, in both cases, there is a unique and directed temporal order of the evolution of each particle taken individually, but there is no absolute time, because there is no unique temporal order of the motion of all the particles taken together, although each particle moves with respect to absolute space. Hence, in this setting, whether or not there is absolute time is an issue of the dynamical structure, by contrast to the question of whether or not there is absolute space.

The drawbacks of this position are the same as the ones of the Newtonian ontology—namely, the commitment to the surplus structure of an absolute space expanding to infinity and, consequently, the impasse into which the subsequent question leads regarding what the particles are, filling space. More precisely, against the background of what we have discussed in Chapter 2, section 1, this impasse amounts to this option being committed to regarding the particles, occupying the points of absolute space, as bare substrata. Hence, this ontology does not meet the requirement of simplicity. It is an illustration of how bringing in more than distance relations individuating matter points creates drawbacks instead of providing additional explanatory value.

(ii) Furthermore, there is a genuine relativistic option for the ontology. This option consists in replacing the distance relation with a distance-temporal relation—that is, a spatio-temporal relation. Instead of matter points that are each separated by a distance from all the other ones, with that distance individuating the matter points, there are continuous lines (i.e. worldlines) with the points on each line being distinguished from the points on each other line by a spatio-temporal interval, which is usually represented in terms of a four-dimensional geometry. All these spatio-temporal intervals exist at once.

This ontology can be cast in terms of a space-time relationalism. No commitment to a space-time that exists in distinction from matter, with matter filling space-time, is called for. In this case, the objections against an absolute space—that is, the objections against the first option—do not apply. By contrast to the relationalist ontology pursued in this book, according to this ontology, instead of matter points being substances that are individuated by distance relations, with these relations changing, there are continuous sequences of matter point events, forming continuous lines (worldlines) that are individuated by the spatio-temporal intervals between them. These intervals do not change. They all exist at once. That is why this ontology is known as *block universe*: all the events throughout the history of the universe exist at once. Again, we stress that the block universe does not have to be conceived in terms of a four-dimensional space-time existing and being filled with matter. It can also be set out in terms of all the spatio-temporal relations between matter point events existing at once.

The main challenge for this ontology is to distinguish between variation within a given configuration and change of that configuration. If one replaces point particles with point events forming continuous lines and a three-geometry with a four-geometry to represent the relations between them, one obtains variation, but no change: the relations represented in terms of a four-geometry provide for variation within the block universe. However, since these relations exist all at once, there is no change. Only when we cut three-dimensional slices through the block and compare them, we can define change in terms of the differences between such slices. But this change concerns only an—arbitrary—description and not the ontology, since the ontology is the four-dimensional block (Geach (1965), in particular, p. 323, is the *locus classicus* of this objection).

To include change, one, therefore, has to stipulate the following over and above the global geometrical order characterizing the block universe (whatever the number of dimensions may be in which that geometry is formulated): the points on each worldline are ordered according to earlier and later, with that order being unique and directed. Then there is a local time for each object (worldline), but no global time. Comparing different points on each worldline, one can introduce change in terms of different spatio-temporal intervals between different points on two worldlines. In this way, the spatio-temporal relations provide for variation and change at once against the background of the points on each worldline being ordered according to earlier and later.

However, in this case, temporal order has to be presupposed as a primitive—that is, the ordering of the points on a worldline according to earlier and later. That ordering obtains as a primitive matter of fact, independently of whether or not there is change in the spatio-temporal intervals between these two lines. Hence, temporal order is not derived from change, but change is derived from the order of the points on each worldline being temporal as a primitive matter of fact. What is the difference between ordering the points on such a line according to, say, below and above and ordering them according to earlier and later? That difference is primitive. Consequently, one cannot do without endorsing a primitive temporal, metrical requirement on the block universe view, even if one casts the block universe view in relationalist terms instead of endorsing an absolute space-time: well-defined spatio-temporal intervals between non-simultaneous point events have to be presupposed as primitive. By contrast, the ontology of matter point substances individuated by distance relations does not rely on any primitive metrical requirement regarding the change of the distance relations—in fact, there is no primitive metrical requirement in this ontology apart from the triangle inequality figuring in the definition of the distance relation (see definition 1 in Chapter 2, section 1).

The question then is whether endorsing a primitive temporal order of the points on each worldline is sufficient for empirical adequacy. There is no possibility to introduce becoming and the passage of time: there is a local temporal order of the points on each world line, but since these points exist all at once, there is no such thing as local becoming (see by contrast Dieks (2006)). The points on any such worldline are ordered according to earlier and later as a primitive matter of fact, but they all exist at once. Becoming and the passage of time then are merely occurring in our experience of the world. As Weyl (1949, p. 116) famously put it, "The objective world simply *is*, it does not *happen*". Consequently, it is doubtful whether this is a conception of change that stands up to scrutiny. In other words, it is doubtful whether paying the price of a primitive temporal order is worthwhile, since doing so arguably does not give us the phenomenology of time. These brief considerations are not intended to be conclusive (see Skow (2015) for a recent defense of the block universe view). Their purpose is to underline again that the ontology of the natural world is a matter of philosophical argument and cannot be settled by referring to the mathematical structure of a physical theory.

(iii) Let us therefore consider how the ontology pursued in the first three chapters of this book fares when it comes to relativistic physics. As stated in the first axiom, there are matter points individuated by the distance relations among them. This ontology thus implements a primitive spatiality requirement in the guise of distances between points. Furthermore, as stated in the second axiom, these distances change, while the matter points are permanent. Since a change in any one distance relation implies a change in all of them (the whole configuration), the order of that change is unique and exhibits a direction. However, there is no primitive temporal requirement. Any measure of that change requires picking out a sub-configuration of distance relations relative to which change is measured.

This hence is an important advantage of this conception: it does not presuppose any primitive temporal requirement. More precisely, it trades a primitive spatiality requirement (distances) with primitive change of these distances for a primitive temporal requirement. The physical theories of relativity then simply bring out clearly the consequences of this move that are already there for any physical theory, from classical mechanics on: it is crucial for this position to distinguish between the ontology on the one hand and the geometry and dynamics on the other. The latter come in as a package to represent the change in the distance relations in a manner that is most simple and informative. In classical mechanics of gravitation, the geometry of Euclidean space and force laws built on attributing parameters to the matter points such as inertial and gravitational mass constitute that package. In physical theories of relativity from classical electrodynamics on, the geometry of (flat or curved) four-dimensional space-time and field laws or laws formulated in terms of direct retarded and advanced particle interaction, both based on attributing parameters to the matter points such as inertial mass and charge, constitute that package. As there is no three-dimensional Euclidean space out there in the world and the matter points do not have properties such as mass or charge *per se*, so there is no (flat or curved) four-dimensional space-time out there in the world.

The package of geometry and dynamics is only our means to represent the variation and the change in the world in the most simple and informative manner. Of course, we have access to the world only through our means of representation. However, it is inappropriate to read off the ontology from the means of representation. Doing so ignores their function—namely, to yield a simple and informative representation and to employ whatever variables are suitable for this purpose. In other words, the justification for these variables (geometrical as well as dynamical ones) is that they are a convenient means to achieve this aim, not that they are what there is out there in the world.

The standard for the ontology is simplicity together with empirical adequacy. Thus, the argument for the ontology of matter points individuated by distance relations that change is its simplicity together with its empirical adequacy. The argument for its simplicity in the context of comparing this ontology to the four-dimensional, spatio-temporal ontology of a block universe (which implements not only a relativistic dynamics but also a relativistic ontology) is that this ontology does not rely on any primitive temporal requirement. The argument for its empirical adequacy in comparison to that ontology is that it has no problem of accounting for the phenomenology of time (becoming, passage of time), because it endorses change in the fundamental relations as a primitive. Admitting change as primitive—but doing so without endorsing a primitive metrical requirement for that change—just is what enables this ontology to accommodate temporal becoming and the passage of time. In a nutshell, this ontology achieves the best of two worlds: the Parmenedian world of eternal being and the Heraclitean world of change. The matter points are substances because they are permanent; they do not come into being, and they do not go out of being. But they are individuated by relations that change as a primitive matter of fact.

If one puts this ontology into the framework of terms that are common in contemporary analytic metaphysics, the substances—the matter points—are *endurants* because they persist without having temporal parts (but they do not have spatial parts either; they do not have any parts at all). Furthermore, since what exists is a spatial configuration of these substances constituted by the distances among them, this is an ontology of *presentism*. However, employing these terms is quite misleading, since they presuppose time as a primitive. In this ontology, there is no time, but only change, and change is what enables this ontology to give an account of the phenomenology of becoming and the passage of time. What exists is a configuration of matter points individuated by distance relations that change. That change exists, but not a whole ordered stack of configurations of matter points— this only is a manner of representation of change. There only is one configuration of matter points of the universe, with the relations that individuate the elements of that configuration changing. This is the only reason why this ontology is akin to presentism. In any case, presentism, thus conceived, is the most simple and parsimonious ontology, since only one configuration exists.

5.4   Super-Humeanism for general relativistic physics

Against this background, let us now apply Leibnizian relationalism including Super-Humeanism to general relativistic physics. Whereas the Super-Humean strategy has been tried out for classical and quantum mechanics in the recent literature, it has as yet not been considered for general relativity theory (GR). The metric formulation of GR involves a set of tensor-fields defined over a four-dimensional semi-Riemannian manifold. Using a local (i.e. component-based) language, we can cast the field equations of the theory in the following form (κ being an appropriate constant):

$$G_{ij}[g_{ij}] = \kappa T_{ij}[\varphi, g_{ij}]. \tag{5.15}$$

For the sake of simplicity, we omit the term involving the cosmological constant. The indices i and j range from 0 to 3. The left hand side of (5.15), being dependent on the metric tensor g_{ij}, contains information about the geometry of the manifold, while the right-hand side conveys information about the physical properties (e.g. energy density, momentum flux) of the material sources mathematically modeled by the portmanteau ϕ, such as the electromagnetic field.

This formulation makes clear that Leibnizian relationalism combined with the Super-Humean strategy faces what appears to be two knock down objections, which we have addressed in general terms in the two preceding sections; we shall now rebut them insofar as they concern GR in particular: in the first place, *space-time in GR is inherently four-dimensional*. There is no objective—that is, non-arbitrary—way to distinguish spatial from temporal relations holding between events. This in turn implies that there is no well-defined notion of spatial configuration in a Leibnizian sense, which undermines the possibility to recover time in the way Huggett (2006) does when applying Super-Humeanism to Newtonian mechanics. Moreover, *GR is considered as the paradigm example of a field theory*. Accordingly, insofar as there is a relationalism worked out for GR in the literature, it is a relationalism that replaces spatial with spatio-temporal relations and particles with fields (see Rovelli (1997) for a prominent example as well as Dieks (2001), section 6; see Pooley (2001) and Pooley (2013), section 7, for a philosophical assessment). The famous hole argument, going back in its contemporary form to Earman and Norton (1987), can be seen as lending further support to such a relationalism.

Following this field relationalism, there is a plenum of fields instead of distance relations between sparse points. However, one may wonder whether this relationalism is worth its name, because there no longer is a principled difference with substantivalism. The reason is the ambiguous status of the metrical field in GR: the metrical field implements the geometry of space-time, but it also carries energy and momentum. For the substantivalist, the metrical field has a special status, being space-time, since it contains the geometry of the universe. For the field relationalist, it is a field interacting with other fields. Both endorse the metrical field as an entity *sui generis*, the substantivalist calling it "space-time", the field relationalist regarding it as a field among other fields. In any case, therefore, this field relationalism relies on a huge amount of geometrical structure, since the metrical field is part of the ontology, even if that field is considered as consisting in metrical relations between point events. Not only is this field ontology thereby committed to primitive geometrical facts, it also gives up the simplicity and coherence of a relationalism that is based only on distance relations: if fields are what there is, they are not properties of anything, but have to be conceived as some sort of a primitive, extended stuff. However, as pointed out in section 2 as well as before in Chapter 2, end of section 1, this implies the commitment to a primitive stuff-essence of matter qua fields in contrast to the individuation of matter qua sparse, unextended points through distance relations on the sparse relationalism proposed in this book.

The relationalist can resist the move to such a field relationalism: in GR as in any other field theory, fields are tested by the motion of particles. There is no direct evidence of fields. All the evidence is one of particle motion. Thus, all empirical determinations of the gravitational field amount to observations of the motion of bodies in the sense of change in the instantaneous spatial relations they stand in. As Einstein puts it,

> The gravitational field manifests itself in the motion of bodies. Therefore the problem of determining the motion of such bodies from the field equations alone is of fundamental importance.
>
> (Einstein and Infeld (1949), p. 209)

Consequently, the field equations are there to determine the motion of bodies. This opens the door for applying the Super-Humean strategy also to GR: instead of buying into the dualism of gravitational field and material bodies as suggested by Einstein and Infeld in this quotation, one can maintain that the ontological bedrock—the Humean mosaic—consists in the motion of bodies only. The gravitational field is a mere representational means that enables us to describe the overall motion of the bodies in a manner that optimizes simplicity and information about that motion.

In practice, though, GR is almost never used as a "closed" theory of interacting point particles. One reason is that, very much like in classical electrodynamics, the N-body problem is ill-defined in GR, because the space-time metric is singular along trajectories of point particles. Another, more pragmatic reason is that on astronomical or even cosmological scales, on which the theory is most often used, the ripples in the space-time produced by individual particles are much too fine grained to be observationally relevant. Instead, the matter content of the relevant systems (be it a galaxy or the entire universe) is usually described in terms of continuous fields, with the energy-momentum tensor entering the field equations being that of an "incompressible fluid" or "cosmic dust". In the literature, this is often interpreted as entailing an ontological commitment to fields rather than particles. Such an interpretation, however, mistakes an effective description for a fundamental one. One can with good reason maintain that the ontological commitment of GR is still with point particles. The relationship between the fundamental particle description and the effective field description (e.g. in terms of fluid dynamics) is one of coarse-graining; cf. our remark on Vlasov and fluid dynamics in the framework of classical mechanics at the end of section 3.1.

This reasoning also applies to the account of purely gravitational phenomena, such as space-time singularities or gravitational waves. Also in these cases all that is physically observable is the change of spatial relations among material bodies. The metric field carrying the information of the configuration of black holes or the propagating gravitational waves is a convenient mathematical means for the dynamical description, which has a very similar role as the electromagnetic field in the theory of classical electrodynamics. However, the fact that the mathematical description makes use of a field by no means implies or even requires the commitment to a field in the ontology, if in the end all that has to be accounted for is the change of spatial relations among matter points. For example, the LIGO experiment setup that detected gravitational waves in 2016 is a precision measurement of a 4 km distance by means of laser interferometry that is able to detect changes of length on the order of one-thousandth the charge diameter of a proton. What was monitored was a certain type of change of a spatial relation between matter points, which can mathematically be conveniently described by a wave rippling through in the metric field.

For simplicity, in the following, we will focus on "pure gravity"—that is, neglect non-gravitational forces—such that all particle motion is geodesic motion according to GR. Again, as mentioned in Chapter 2, section 1 in general terms, this Leibnizian stance cannot recognize all mathematically possible solutions of in this case the field equations of GR as describing physically possible worlds. In particular, (5.15) allows for empty cosmological solutions—that is, models in which the universe is totally deprived of matter—yet space-time has, say, ripples and lumps. These solutions have to be dismissed as mathematical surplus structure of the theory; taking them ontologically serious would amount to inflating the ontology with the gravitational field as a substance *sui generis* existing over and above matter. Again, dismissing these solutions is no problem, since the world we live in undoubtedly is not an empty universe. By the same token, also the solutions lacking (global) Cauchy hypersurfaces—that is, three-surfaces that are intersected only once by any non-space-like curve—have to be dismissed, since these solutions would depict cosmological models in which it is even not possible to give a physical meaning to a space/time split.

Although all these solutions can be dismissed as mathematical surplus without ontological significance and although the objection from GR being a field theory can be countered on the basis of all empirical evidence consisting in the motion of bodies, the fact remains that the fundamental relations in GR seem to be spatio-temporal ones between events rather than distance relations between particles that change. This comes out clearly even if one focuses only on models of space-times that admit space-like Cauchy surfaces. Consider a solution of (5.15) consisting in a four-geometry that includes word-lines of material bodies, such that the manifold possesses a product topology $\Sigma_3 \times \mathbb{R}$. Then, we can reduce the physical description of this four-geometry to the sum of descriptions on a pile of three-dimensional Cauchy surfaces Σ_3 that cut the manifold in the space-like direction. The manifold can thus be foliated by a series of space-like leaves. Nonetheless, the inherent four-dimensionality of GR shows up in the fact that the choice of the foliation is not unique; for any choice of three-foliations, we always recover the same four-dimensional physical representation including the worldlines of the material bodies.

Again, however, the arguments spelled out in the preceding section suggest taking the four-geometry as a means of representation instead of endorsing it as constituting the ontology of GR in the guise of a block universe. It is true that the distances making up the configuration of matter points and their change cannot be specified uniquely in GR. However, this just shows that fundamental space-like facts can be *described* in different yet equivalently simple and strong ways. We should not conclude from this descriptive underdetermination that there are no fundamental space-like, relationalist facts, on pain of losing the simplicity, coherence and empirical adequacy of the ontology.

This situation is in a certain sense analogous to the one of Super-Humeanism in the area of quantum mechanics: on Super-Humeanism applied to Bohmian mechanics, for instance, the particle positions and their change make up the Humean mosaic. However, due to the universe being in quantum equilibrium, our knowledge is limited to what can be obtained by applying Born's rule. Furthermore, there are two different formulations of Bohmian mechanics—the standard one and the identity-based one—that agree on the accessible facts, but disagree on the particle trajectories (see Chapter 3, section 2). Thus, the Humean mosaic cannot be uniquely specified in the Bohmian primitive ontology approach. In general, a limit of accessibility applies to any primitive ontology proposed for quantum mechanics, as pointed out at the end of Chapter 3. Nonetheless, the argument for these theories is that they solve the measurement problem by providing an ontology and a dynamics that make unique measurement outcomes intelligible. Hence, the issue of what is the Humean mosaic is one of ontological argument and not whether or not the Humean mosaic can be uniquely described or is accessible to observation.

Recent progress in physics provides a further argument to vindicate the Leibnizian perspective on space and time in general relativistic physics. Gomes et al. (2011) show that there is an alternative theory defined on the phase space of GR. Such a theory is not just alternative, but *dual* to GR in the sense that, under the appropriate choice of gauge, the dynamical trajectories of the two theories coincide. This dual theory is Barbour's shape dynamics (see Gomes and Koslowski (2013) for a concise presentation and see the discussion of Barbour's theory of classical mechanics in Chapter 3, section 1).

Leaving aside technical considerations, the important philosophical consequence of this duality in classical (that is, non-quantum) gravity is that there are two metaphysical stances with respect to space and time that are compatible with the empirical predictions of general relativistic physics. The first one is the irreducibly four-dimensional perspective of GR; the second one is the three-dimensional perspective of shape dynamics. In this theory, dynamics can be depicted as a succession of (conformal) three-geometries. Unlike GR, this theory is *not* invariant under change of foliation. Instead, this characteristic symmetry of GR is traded for a local conformal symmetry, which means that the geometry defined on each folio lacks a privileged notion of scale—it is not *size*, but *shape* that matters.

In shape dynamics, there is a well-defined notion of instantaneous spatial configuration. This fact does not imply that instants are referred to an external clock ticking in the background. What exists is a succession of spatial configurations that can be arbitrarily labeled by a monotonically increasing parameter, but this succession is not *in time*. On the contrary, it defines a "time-like" direction for the unfolding of the dynamics. Hence, shape dynamics is compatible with Leibniz' (and Mach's) view of space as the order of coexistence (i.e. bodies related by spatial relations) and time as a bookkeeping device to describe the order of change (i.e. the succession of instantaneous spatial configurations).

We do not claim that shape dynamics fully implements the ontology of Leibnizian relationalism. First of all, also this theory is a field theory. Secondly, at this stage, the theory has been worked out only for pure gravity (but see Gomes and Koslowski (2012) for some first results for gravity-matter coupling). Thirdly, this theory includes irreducible geometric facts regarding angles (that is, a conformal structure) contra our fundamentally geometry-less ontology. However, as argued in Chapter 3, section 1, it would be a misunderstanding to require that a relationalist ontology has to be vindicated by a physical theory that employs only relationalist representational means. The fact that Barbour's theory, which is the most detailed relationalist physical theory for both classical mechanics and general relativistic physics as yet, has to rely on a primitive conformal structure for the very concept of shape to be meaningful just underlines again that when it comes to a physical theory, more representational means are needed than what is provided by the sparse ontology of distance relations only, although this ontology is fully empirically adequate.

The point at stake here is that, to put it in the terms of Gryb and Thébault (2016), gravity in general relativistic physics is "Janus-faced": it is compatible both with a block universe where there is variation of local matters of fact but no change (and, hence, no time ordering change), and with a dynamical universe where there is change in the guise of change in the fundamental relations that connect the fundamental objects, on which temporal facts supervene in the form of a monotonically increasing parameter labelling the succession. In short, the duality of GR and shape dynamics again shows that one cannot read the ontology off from the mathematical structure of a physical theory, since one would in this case end up with two proposals for the ontology of general relativistic physics that contradict each other, based on two different formalisms for general relativistic physics. The ontology has to be settled by criteria such as parsimony and coherence together with empirical adequacy.

Against this background, taking the distance relations among the point particles and their change throughout the history of the universe to be the Humean mosaic, let us now exploit for GR the Super-Humean strategy that Huggett (2006) developed in order to vindicate a Leibnizian relationalist ontology for Newtonian mechanics. To briefly recall (see Chapter 3, section 1, for details), Huggett (2006) employs the concept of adapted frame, that is, a reference frame tied to a body (e.g. a material particle)— otherwise said, an assignment of N-tuples of real numbers over time such that the origin $(0, 0, \ldots, 0)$ corresponds to the body to which the frame is adapted. Inertial frames then are those frames in which an observer would describe the history of relations in the simplest and strongest way in terms of Newton's laws. Note that, in order to exploit this characterization, we do not necessarily need any body to be actually tied to an inertial frame. This is because, once we have a class of adapted frames, unoccupied frames can be defined as those related to the occupied ones by means of a spatial rigid translation. As pointed out in Chapter 3, section 1, the notions of adapted frame and spatial translation do not require any ontological commitment over and above what is included in axioms 1 and 2: they just presuppose matter points and change of spatial relations among them. Once this characterization of inertial frame is in place, absolute acceleration can be reduced to the history of change of the spatial relations holding between an inertial and a non-inertial frame: acceleration is part of the description of how much the pattern that constitutes the history of a non-inertial frame deviates from the regularities encoded in the inertial pattern as seen in an inertial frame. By the same token, any other absolute quantity of motion such as, for example, angular momentum, can be shown to supervene on the Humean mosaic.

When it comes to GR, as in the classical case, we can define an adapted frame of reference as one in which a certain particle is always at rest at the origin, so that all distances along the axes are just the distances from that particle (i.e. the arbitrary way in which we label spatial relations). We can then relate all reference frames by means of rigid spatial translations (i.e. continuous functions that send N-tuples of coordinates in one frame to N-tuples of coordinates in another frame in a way that preserves our criterion to assign labels to spatial relations). We look as usual for regularities in the history that admit a particularly simple yet strongly informative description. However, the problem that we face in this respect is that there is no such thing as classical inertial motion in GR: this theory merges inertial structure and the gravitational field. Consequently, there is no unique way to separate them, let alone find a case where gravitational effects can be neglected in a region bigger than a point. Hence, the description that we search for cannot be encoded in a law of the form $\ddot{x}^i = 0$ (a dot indicates the usual differentiation with respect to an arbitrary time-like parameter defined along a given pattern); in this case, we would be dealing with a Newtonian world rather than a general relativistic one.

Nonetheless, even if we cannot make use of the concept of inertial motion in GR, still we can employ a generalization of this concept: the simplest and most informative description of the regularities in a general relativistic world is that of geodesic motion (more precisely, non-space-like geodesic motion)—that is, $\ddot{x}^i + \Gamma^i_{jk}\dot{x}^j\dot{x}^k = 0$, using an appropriate (affine) parametrization. Geodesic motion is a well-defined concept in GR, since the theory stipulates that the motion of freely-falling bodies follows geodesic trajectories. Here we immediately see that the (Levi-Civita) connection with coefficients Γ^i_{jk} is not a geometrical feature crafted so to speak in the Humean mosaic: it is just a tool that helps to formalize how the description will change among two different observers moving geodesically—in other words, a description of how two frames adapted to freely-falling bodies and related by a rigid spatial translation differ.

Note that we do not presuppose the existence of a space-time structure (in particular, a connection) that *defines* what it is for a motion to be geodesic, but, rather, the other way round: we define geodesic motion as a particularly simple pattern in the history of relational change. Then we construct a connection as a bookkeeping device that accounts for the relational differences among different geodesics: there are no primitive geometrical facts, just descriptive tools taken from the language of (differential) geometry. Similar to the classical case, here there might actually be no body moving geodesically: what we require is just that any other adapted frame can be related to a geodesic one by means of a spatial rigid translation. If this cannot be achieved, then the possible world we are dealing with does not admit the laws of general relativistic physics as theorems of the simplest and strongest system.

It is worth noting that, in the Newtonian case, the coefficients of the connection do not arise, because all inertial motions are identical, which is consistent with the fact that Newtonian space-time admits a flat connection (i. e. the simplest and strongest description of inertial motion is one for which $\Gamma^i_{jk} = 0$). In the general relativistic case, by contrast, geodesic motions might in fact differ even when adopting the same affine parametrization (what is usually called "geodesic deviation", which is at the root of the explanation of how tidal forces arise). From this point on, all the relevant geometrical and dynamical features of the theory can be shown to supervene on the Humean mosaic in the usual manner. For example, curvature is a simple and informative way to describe relational change between freely-falling bodies; more generally, the metrical structure g_{ij} of space-time supervenes on geodesic motions through the usual relation $\Gamma^i_{jk} = \frac{1}{2} g^{hi} \left(g_{kh,j} + g_{hj,k} - g_{jk,h} \right)$ (a comma indicates standard differentiation with respect to the subsequent index). Geodesic trajectories also suffice to fix the topology of space-time, as shown in Malament (1977). Note that, by construction, the geometrical description encoded in $\Gamma^\mu_{v\sigma}$ and $g_{\mu v}$ does not inhere within *single* trajectories, but supervenes on the *totality* of geodesic motions. In this sense, the continuous large-scale geometry of space-time results from a coarse-grained description of the history of relational change among particles.

In general, all fields—including, as we have seen, the metrical one—are in this framework shortcuts to describe certain forms of particle motion. Hence, the relationship between particles and fields is not physical—particles do not generate fields, nor are they pushed by fields—but descriptive: a particle-vocabulary can be coarse grained to a field one in order to simplify the description of gravitational phenomena without loss of physical information. Then, quantities such as momentum densities, energy flows and the like can be constructed out of this field-vocabulary. This issue is crucial: if the ontology is the one of Leibnizian relationalism, quantities such as the stress-energy tensor do not refer directly to the underlying particle motions, but express their field-like description; they are, so to speak, a description of how a description changes. Clarifying this point is very important because if we took particles and fields on a par and, in particular, regarded particles as field sources, then singularities would arise. This is all the more true for GR. In this theory, a Schwarzschild radius is associated to any extended body, which depends on the body's mass: roughly speaking, if the body's extension drops below this radius, then the body collapses into a black hole singularity. It is then obvious why, in GR, this notion of massive point-particle makes little sense besides some particular approximations. However, it should be clear that our notion of particle is *not* this notion of massive point-particle.

What has been said so far shows that there is nothing strange in claiming that a large-scale continuum theory such as GR (i.e. a theory employed to model continuous material systems at astrophysical or cosmological scales) admits an extremely sparse ontology in terms of distance relations individuating matter points. Nonetheless, these results do not show that the central law of GR—namely, the field equations (5.15)—can be recovered from the Humean mosaic as a theorem of the system that strikes the best balance between simplicity and strength in describing the change in the distance relations among the point particles that occurs throughout the history of the actual world. Indeed, thus recovering the field equations (5.15) of GR is outright impossible for the following reason: if the Humean mosaic are distance relations among sparse matter points and their change, we can obtain geometry and dynamics in a package as the system that optimizes simplicity and information in the description of that change. However, geometry and dynamics remain distinct, as in the classical case. In other words, the dynamical laws that describe such a Humean mosaic are those of a theory formulated over a fixed (albeit non-Euclidean) background. There is nothing in the regularities of motion of such a Humean mosaic that could lead to a higher order description where geometry is *coupled* to matter in a dynamical way. The ontology of there being only distance relations and their change is too meagre to get (5.15) from such a possible world by Humean means: the most we can get is a possibly very complex, but still non-dynamical geometry compatible with the geodesic law of motion. This is because geometry supervenes on the mosaic by means of the law of geodesic motion, while non-gravitational fields come out as a way to describe non-geodesic motion: given this construction, there is no way to conjecture any inter-dependence between geometry and material fields.

That notwithstanding, this result can be accommodated in an ontology of there being only distance relations among sparse matter points and their change that does justice to general relativistic physics. Consider the class of cosmological solutions of (5.15) (modulo those that we have previously discarded as mathematical surplus): each of these models describes a possible world in which GR holds. These models are each characterized by a certain metric g_{ij} (compatible with the Levi-Civita connection) and a certain stress-energy tensor T_{ij}. The main difference that this cluster of possible worlds bears with respect to a cluster of Newtonian worlds is that the

Newtonian worlds *always* feature the same spatial and temporal metrical structures. By contrast, in general, $g_{\mu\nu}$ co-varies with $T_{\mu\nu}$ from world to world in the GR case. This is one possible sense in which we can understand what is known as the background independence of GR: the same spatio-temporal structures of Newtonian mechanics appear in all Newtonian models, so that they are nomologically necessary; spatio-temporal structures in GR, by contrast, are nomologically contingent.

Against this background, what (5.15) represents is the manner in which in any possible world of GR, spatio-temporal structures described by some metric tensor g_{ij} are correlated with material structures described by a stress-energy tensor T_{ij}. That is to say, in any possible world of the ontology of distance relations among sparse matter points and their change in which GR is valid, the system that strikes the best balance between simple and being informative about a particular such world is one in which a dynamics describes the evolution of the configuration of matter over a fixed background—that is, a particular cosmological solution of (5.15) describing a particular possible world of GR. The general law (5.15) then is not a theorem that generalizes the regularities within a world, but a *trans-world* generalization: it expresses the relationship between the mathematical structures formulating the geometry and the mathematical structures formulating the dynamics in any possible world where GR is the best system describing the change in the distance relations among the point particles. In sum, we propose a generalization of the Super-Humean account that Huggett (2006) developed for Newtonian mechanics that shows that geometry and dynamics supervene on the Humean mosaic of Leibnizian relations *at each world* by means of a geodesic law; we then obtain Einstein's field equations as the simplest and strongest description of the correlation between geometry and dynamics for the whole class of these worlds.

We consider only *classical* gravitation (in a Newtonian as well as general relativistic setting), but not *quantum* gravity, since, by contrast to the standard model of QFT, there is no established quantum theory of gravitation as yet. The discussion of Leibnizian relationalism applied to general relativistic physics in this section and the elaboration of a (Bohmian) particle ontology for QFT in Chapter 4, both conceived in the Super-Humean setting, complete our case for the minimalist ontology set out in Chapter 2 being able to cover all existing physics, in fact being the best proposal for an ontology of the natural world that is based on our scientific knowledge in fundamental physics as a whole.

Nonetheless, let us add a few sketchy remarks with respect to the claims about the alleged emergence of space-time in a future theory of quantum gravity (see, e.g., Huggett and Wüthrich (2013) for a sympathetic overview and, e.g., Esfeld and Lam (2013) for voicing scepticism):

1. As in any quantum theory, so also in a future theory of quantum gravity, one cannot settle ontological issues on the basis of an operator

formalism. One has to spell out an ontology that solves the measurement problem and thereby establishes the link between the theory and the empirical evidence (which is supposed to confirm the theory).

2. The emergence of time is nothing new and poses no problem. Indeed, the Leibnizian relationalism proposed in this book includes what can be described as the emergence of time by deriving time from change. Thus, the emergence of time can be studied already for classical mechanics. More importantly, Barbour's framework, which includes the emergence of time on the basis of a commitment to fundamental spatial relations, can be put to work also in the domain of quantum gravity (see, e.g., Barbour et al. (2014) and Gryb and Thébault (2016)). Hence, one has to distinguish between the emergence of time and the emergence of space: the former, by contrast to the latter, is well studied.

3. As elaborated on in Chapter 2, section 1, without acknowledging a plurality of objects in some sense—be it so-called thin objects without intrinsic properties (see French (2010))—it is difficult to see how empirical adequacy could be achieved. If there is a plurality of objects, there has to be a certain type of relations in virtue of which these objects make up a configuration that then is the world. This type of relations has furthermore to be such that it characterizes the world as a physical or material world, by contrast to, say, a hypothetical world of Cartesian minds that are individuated by thinking relations. That is the reason to single out a type of relations providing for extension— namely, the spatial relations. These relations are able to constitute a physical or material world by individuating simple objects that then are matter points and to lead to empirically adequate theories through including their change as second axiom in the ontology and formulating a dynamical structure that describes the change in the spatial relations. No one has as yet worked out a theory in which another type of relations (from which spatial relations may then emerge) performs that task.

Of course, there is no *a priori* argument independent of science for spatial relations being the fundamental relations that unify the world. To put it in the spirit of Popper, the ontology set out here is an audacious proposal that is open to falsification. Future science may teach us other lessons. However, instead of simply announcing such lessons on the basis of not yet existing physical theories, one should work out how the criteria (1) and (3) can be met in a way that is not based on fundamental distance relations.

5.5   Relativistic physics and quantum entanglement

Instead of making the effort to argue that Leibnizian relationalism together with Super-Humeanism stands also firm as the best proposal for the ontology of general relativistic physics, one can try to exploit the tension between relativistic physics on the one hand and quantum mechanics and QFT on the other hand in order to vindicate an ontology that is based on spatial relations rather than spatio-temporal ones. That tension is brought by Bell's theorem. This theorem is based on the following two premises:

1. *No conspiracy*: the choice of the variables to be measured on a physical system is independent of the past state of the system.
2. *Locality*: an event can only be influenced by events that are located in its past light cone.

Bell's theorem then proves that no theory that endorses these two premises can reproduce the statistical predictions of measurement outcomes of quantum mechanics, more precisely the statistics of the measurement outcomes on entangled quantum systems (see Bell (2004), in particular, ch. 2, 7 and 24; see, furthermore, Seevinck and Uffink (2011)). Since any future physical theory will have to reproduce the statistical predictions of measurement outcomes of quantum mechanics that are confirmed by experiment, Bell's theorem poses a constraint not only on hidden variable theories of quantum mechanics but also on any future physical theory.

The no conspiracy premise is a general premise that concerns all experimental investigations of nature: if the past state of the system under investigation influenced the choice of the variables to be measured on the system, then no reliable information about the system could be obtained through experimental investigation. This premise is compatible with determinism; it is, for instance, satisfied in Bohmian mechanics (see Esfeld (2015)). Since the no conspiracy premise is a general premise of natural science that is not specific for quantum physics, or relativity physics, the locality premise is the one that Bell's theorem rules out. More precisely, Bell's theorem is widely taken to establish non-locality in the sense that events that are separated by a space-like interval influence each other, such as the events of the measurement settings and measurement outcomes across both wings of the EPR experiment (see the seminal monograph of Maudlin (2011)). Note that such an influence does not automatically mean that there is direct causation between space-like separated events. The correlation between these events can also be accounted for in terms of a common cause, but that common cause then is non-local. Both Bohmian mechanics and the GRW matter density theory can be received as providing such a common cause explanation of the correlated measurement outcomes in the EPR experiment (see Egg and Esfeld (2014)).

Recent research on the tension between quantum non-locality and relativity physics has shown that it is possible to formulate the dynamical structure of quantum physics in the primitive ontology framework in a Lorentz invariant manner if one considers the entire spatio-temporal distribution of the elements of the primitive ontology throughout the history of the universe as a whole; in particular, Tumulka (2006, 2009) has established this result in the framework of the GRWf theory with flash-events as the primitive ontology (see Bedingham et al. (2014) for a similar result for the GRWm theory with a field as the primitive ontology). However, this result does not obtain if one considers the *evolution* of the configuration of the elements of the primitive ontology starting with an arbitrary initial configuration, not to mention the *interaction* between the elements of the primitive ontology: there is no means to represent that evolution (and the interaction) in a Lorentz invariant manner available—that is, without including influences that connect space-like separated events in the dynamics so that a definite temporal order of these events has to be assumed—although that temporal order is not accessible in experiments (see Esfeld and Gisin (2014) and Barrett (2014)). Again, this is a feature of the measurement problem: everything is fine with respect to Lorentz invariance as long as one considers only possible evolutions of a given initial configuration of, say, GRW flashes and assigns probabilities to these possible evolutions. However, one cannot understand the realization of one such evolution (corresponding to the occurrence of particular measurement outcomes) in a Lorentz invariant manner.

By way of consequence, the recent research on Lorentz invariance in the framework of the GRWf and the GRWm theories constitutes no argument to prefer the ontology of flashes or a field with a GRW-type collapse dynamics to the ontology of permanent particles with a Bohmian dynamics. The argument of Chapter 3, section 3, for Bohmian mechanics providing the best solution to the measurement problem stands firm also in view of these recent results. As we briefly mentioned at the end of Chapter 4, the dynamical structure of Bohmian mechanics as well as Bohmian QFT requires the committed to a preferred foliation of flat Minkowski space-time into space-like hypersurfaces, but this foliation does not have to be introduced by hand: it is possible to consider the universal wave function as determining that preferred foliation. Nevertheless, the theory then is Lorentz invariant on the level of empirical predictions. It is not possible to discover that preferred foliation through experiments (see Dürr et al. (2013a) and the brief discussion in Chapter 4, section 6). In any case, problems that primitive ontology theories of quantum physics may face with respect to the dynamical structure of special and GR theory cannot be counted as an argument against these theories: they solve the measurement problem. No one has produced a solution to this problem that (a) acknowledges determinate measurement outcomes and (b) is a relativistic theory of interactions, including in particular measurement interactions (see again Barrett (2014)).

These results as well as the general argument for the tension between quantum non-locality and relativity physics are conceived in the framework of a field formulation of the dynamics and including the background assumption that advanced action is ruled out. The field formulation of classical electrodynamics situates every influence in the immediate past of a given event by representing interaction as being transmitted by fields. Accordingly, we have formulated the locality premise that enters Bell's theorem in such a way that locality means local influences being situated in the past light cone of a given event. Bell's theorem then establishes that quantum physics violates this locality principle (if one takes the general no conspiracy premise for granted). Against the background of a field formulation of the dynamics, it is obvious why this locality principle breaks down in the quantum case: the wave function is a field parameter. However, it is not a field parameter on ordinary space, but on configuration space, being a field following a local dynamics only on configuration space. By way of consequence, in contrast to relativity physics, the wave function correlates the motion of spatially distant particles (and does so independently of their distance in physical space).

Nonetheless, as in the case of classical electrodynamics, so also in the quantum case, one can conceive a formulation of the dynamical structure that is based on both retarded and advanced particle interaction (see Cramer (2016) for the most prominent physical model and notably Price (1996), ch. 9, Dowe (1996) and Corry (2015) for philosophical endorsements; there even is a version of Bohmian mechanics with advanced action, see Sutherland (2008)). Generally speaking, whenever there is a dynamical structure of a physical theory in terms of direct interaction between spacelike separated events, one can in principle also conceive a structure that transforms that direct interaction into an advanced action of future events. By contrast to the case of classical electromagnetism as illustrated by the Wheeler-Feynman theory of direct particle interactions, in the quantum case, the advanced action option, however, has to include the field parameter, working with both retarded and advanced waves in the dynamical structure. That difference notwithstanding, by buying into advanced action, one avoids quantum non-locality in the sense of influences that connect space-like separated events (be it in terms of direct causation, be it in terms of a non-local common cause). In the case of advanced action, all influences are represented in terms of waves that go exclusively through the light cones, but in addition to influences from the past light cone on a given event, there also is an influence coming from the future light cone of the event in question.

Consequently, by taking the advanced action option, one cannot restore locality in the sense of the locality premise formulated above. There is no question of a dynamical structure that situates every influence on a given event in the (immediate) past light cone of the event in question. Admitting advanced action infringes furthermore upon the no conspiracy premise: there is no independence between the past state of the system and the choice of the measured parameter. However, there is no direct influence from the past state of the system on the choice of the measured parameter.

Both are correlated only through the influence from the future measurement outcomes. Any advanced action theory entails such a correlation. This correlation established through the influence from the future does not necessarily result in a paradox or a conspiracy (see, e.g., Price (1996), ch. 9, and Lazarovici (2014)). In brief, a direct interaction dynamics can in principle be empirically adequate. It provides a relativistic dynamics also for the quantum case, albeit one that includes advanced in addition to retarded action.

The in principle availability of a dynamical structure in terms of both retarded and advanced action in the case of relativity physics at least as far as classical electromagnetism is concerned as well as in the quantum case at least as far as quantum mechanics is concerned shows that there is not necessarily a conflict between the dynamical structures of relativity physics and quantum physics. By way of consequence, it is short-sighted to base an argument for an ontology of spatial rather than spatio-temporal relations on the tension between quantum physics and relativity.

The discussion of relativistic physics rather underlines again the need to distinguish between the ontology of the natural world and the dynamical structure used in the formulation of a physical theory: there is no point in reading the ontology off from the dynamical structure that one chooses. We need criteria for the ontology that are independent of considerations of the dynamical structure—that is, criteria that are not touched by the underdetermination of dynamical structure and the change of dynamical structure in the history of physics. Parsimony—together with empirical adequacy—is the central criterion for ontology, because one has to justify why one should admit an entity to the ontology, and its figuring in the dynamical structure of a physical theory does not provide such a justification. Parsimony leads to the question of what is an ontology of the natural world that is minimally sufficient for empirical adequacy. Leibnizian relationalism combined with Super-Humeanism—that is, an ontology of distance relations individuating matter points and the change of these relations—is the answer to this question. As set out in this book, we thus achieve an ontology that covers all known physics while being most parsimonious.

Bibliography

Albert, D. Z. (2015). *After physics*. Cambridge, Massachusetts: Harvard University Press.

Albert, D. Z. and Loewer, B. (1996). Tails of Schrödinger's cat. In Clifton, R. K., editor, *Perspectives on quantum reality*, pages 81–91. Dordrecht: Kluwer.

Allori, V., Goldstein, S., Tumulka, R., and Zanghì, N. (2008). On the common structure of Bohmian mechanics and the Ghirardi-Rimini-Weber theory. *British Journal for the Philosophy of Science*, 59(3):353–389.

Allori, V., Goldstein, S., Tumulka, R., and Zanghì, N. (2014). Predictions and primitive ontology in quantum foundations: a study of examples. *British Journal for the Philosophy of Science*, 65(2):323–352.

Anderson, E. (2013). The problem of time and quantum cosmology in the relational particle mechanics arena. *arXiv:1111.1472v3 [gr-qc]*.

Anderson, E. (2015). Configuration spaces in fundamental physics. *arXiv:1503. 01507v2 [gr-qc]*.

Ariew, R., editor (2000). *G. W. Leibniz and S. Clarke: Correspondence*. Indianapolis: Hackett.

Armstrong, D. (2004). Going through the open door again: counterfactual versus singularist theories of causation. In Collins, J., Hall, N., and Paul, L. A., editors, *Causation and counterfactuals*, pages 445–457. Cambridge, Massachusetts: MIT Press.

Arntzenius, F. (1994). Electromagnetic arrows of time. In Horowitz, T. and Janis, A., editors, *Scientific failure*. Rowman & Littlefield, Maryland.

Arntzenius, F. and Hawthorne, J. (2005). Gunk and continuous variation. *The Monist*, 88:441–465.

Baker, D. J. (2009). Against field interpretations of quantum field theory. *British Journal for the Philosophy of Science*, 60:585–609.

Barbour, J. (2003). Scale-invariant gravity: particle dynamics. *Classical and Quantum Gravity*, 20:1543–1570.

Barbour, J. (2012). Shape dynamics. An introduction. In Finster, F., Müller, O., Nardmann, M., Tolksdorf, J., and Zeidler, E., editors, *Quantum field theory and gravity*, pages 257–297. Basel: Birkhäuser.

Barbour, J. and Bertotti, B. (1982). Mach's principle and the structure of dynamical theories. *Proceedings of the Royal Society A*, 382:295–306.

Barbour, J., Koslowski, T., and Mercati, F. (2014). The solution to the problem of time in shape dynamics. *Classical and Quantum Gravity*, 31(15):155001.

Barrett, J. A. (2014). Entanglement and disentanglement in relativistic quantum mechanics. *Studies in History and Philosophy of Modern Physics*, 48:168–174.

Bauer, G. (1997). *Ein Existenzsatz für die Wheeler-Feynman-Elektrodynamik*. München: Herbert Utz Verlag.

Bauer, G., Deckert, D. A., and Dürr, D. (2013). On the existence of dynamics in Wheeler-Feynman electromagnetism. *Zeitschrift für angewandte Mathematik und Physik*, 64(4):1087–1124.

Bauer, G., Deckert, D.-A., Dürr, D., and Hinrichs, G. (2014). On irreversibility and radiation in classical electrodynamics of point particles. *Journal of Statistical Physics*, 154(1):610–622.

Bauer, G., Deckert, D.-A., Dürr, D., and Hinrichs, G. (2016). Global solutions to the electrodynamic two-body problem on a straight line. *arXiv:1603.05115* [math-ph].

Baumgartner, M. (2013). A regularity theoretic approach to actual causation. *Erkenntnis*, 78:85–109.

Bedingham, D., Dürr, D., Ghirardi, G. C., Goldstein, S., Tumulka, R., and Zanghì, N. (2014). Matter density and relativistic models of wave function collapse. *Journal of Statistical Physics*, 154:623–631.

Beebee, H. (2000). The non-governing conception of laws of nature. *Philosophy and Phenomenological Research*, 61:571–594.

Beebee, H. (2006). Does anything hold the universe together? *Synthese*, 149:509–533.

Beebee, H. and Mele, A. (2002). Humean compatibilism. *Mind*, 111:201–223.

Bell, J. S. (2004). *Speakable and unspeakable in quantum mechanics. Second edition.* Cambridge: Cambridge University Press.

Belot, G. (1999). Rehabilitating relationalism. *International Studies in the Philosophy of Science*, 13:35–52.

Belot, G. (2000). Geometry and motion. *British Journal for the Philosophy of Science*, 51(4):561–595.

Belot, G. (2001). The principle of sufficient reason. *Journal of Philosophy*, 98:55–74.

Belot, G. (2011). *Geometric possibility*. Oxford: Oxford University Press.

Belot, G. (2012). Quantum states for primitive ontologists: a case study. *European Journal for Philosophy of Science*, 2(1):67–83.

Belousek, D. W. (2003). Formalism, ontology and methodology in Bohmian mechanics. *Foundations of Science*, 8:109–172.

Benedikter, N., Porta, M., and Schlein, B. (2014). Mean field evolution of fermionic systems. *Communications in Mathematical Physics*, 331(3):1087–1131.

Bhogal, H. and Perry, Z. R. (2017). What the Humean should say about entanglement. *Noûs*, 51(1):74–94.

Bigaj, T. (2015). Dissecting weak discernibility of quanta. *Studies in History and Philosophy of Modern Physics*, 50:43–53.

Bird, A. (2007). *Nature's metaphysics: laws and properties*. New York: Oxford University Press.

Black, R. (2000). Against quidditism. *Australasian Journal of Philosophy*, 78:87–104.

Blackburn, S. (1990). Filling in space. *Analysis*, 50:62–65.

Bohm, D. (1951). *Quantum theory*. Englewood Cliffs: Prentice-Hall.

Bohm, D. (1952a). A suggested interpretation of the quantum theory in terms of "hidden" variables. 1. *Physical Review*, 85(2):166–179.

Bohm, D. (1952b). A suggested interpretation of the quantum theory in terms of "hidden" variables. 2. *Physical Review*, 85(2):180–193.

Bohm, D. and Hiley, B. J. (1993). *The undivided universe. An ontological interpretation of quantum theory*. London: Routledge.

Bohm, D., Hiley, B. J., and Kaloyerou, P. N. (1987). A causal interpretation of quantum fields. *Physics Reports*, 144:349–375.

Boltzmann, L. (1896). *Vorlesungen über Gastheorie*. Leipzig: Barth.

Briceno, S. and Mumford, S. (2016). Relations all the way down? Against ontic structural realism. In Marmodoro, A. and Yates, D., editors, *The metaphysics of relations*, pages 198–217. Oxford: Oxford University Press.

Brown, H. R., Dewdney, C., and Horton, G. (1995). Bohm particles and their detection in the light of neutron interferometry. *Foundations of Physics*, 25(2):329–347.

Brown, H. R., Elby, A., and Weingard, R. (1996). Cause and effect in the pilot-wave interpretation of quantum mechanics. In Cushing, J. T., Fine, A., and Goldstein, S., editors, *Bohmian mechanics and quantum theory: an appraisal*, volume 184 of *Boston Studies in the Philosophy of Science*, pages 309–319. Dordrecht: Springer.

Butterfield, J. (2006a). Against pointillisme about geometry. In Stadler, F. and Stölzner, M., editors, *Time and history. Proceedings of the 28th International Ludwig Wittgenstein Symposium*, pages 181–222. Frankfurt (Main): Ontos.

Butterfield, J. (2006b). Against pointillisme about mechanics. *British Journal for the Philosophy of Science*, 57:709–753.

Callender, C. (2007). The emergence and interpretation of probability in Bohmian mechanics. *Studies in History and Philosophy of Modern Physics*, 38:351–370.

Callender, C. (2011). Philosophy of science and metaphysics. In French, S. and Saatsi, J., editors, *The Continuum companion to the philosophy of science*, pages 33–54. London: Continuum.

Callender, C. (2015). One world, one beable. *Synthese*, 192(10):3153–3177.

Castañeda, H.-N. (1980). Causes, energy and constant conjunctions. In van Inwagen, P., editor, *Time and cause. Essays presented to Richard Taylor*, pages 81–108. Dordrecht: Reidel.

Cei, A. and French, S. (2014). Getting away from governance: a structuralist approach to laws and symmetries. *Methode*, 4:25–48.

Chakravartty, A. (2013). On the prospects of naturalized metaphysics. In Ross, D., Ladyman, J., and Kincaid, H., editors, *Scientific metaphysics*, pages 27–50. Oxford: Oxford University Press.

Cohen, J. and Callender, C. (2009). A better best system account of lawhood. *Philosophical Studies*, 145(1):1–34.

Colin, S. (2003). Beables for quantum electrodynamics. *Fondation Louis de Broglie. Annales*, 29(1–2):273–296.

Colin, S. and Struyve, W. (2007). A Dirac sea pilot-wave model for quantum field theory. *Journal of Physics A*, 40(26):7309–7341.

Corry, R. (2015). Retrocausal models for EPR. *Studies in History and Philosophy of Modern Physics*, 49:1–9.

Cowan, C. W. and Tumulka, R. (2016). Epistemology of wave function collapse in quantum physics. *British Journal for the Philosophy of Science*, 67:405–434.

Cramer, J. G. (2016). *The quantum handshake. Entanglement, nonlocality and transactions*. Cham: Springer.

Curceanu, C. and alteri (2016). Spontaneously emitted x-rays: an experimental signature of the dynamical reduction models. *Foundations of Physics*, 46(3):263–268.

Davidson, D. (1995). The problem of objectivity. *Tijdschrift voor Filosofie*, 57:203–220.

Dawid, R. and Thébault, K. P. Y. (2014). Against the empirical viability of the Deutsch-Wallace-Everett approach to quantum mechanics. *Studies in History and Philosophy of Modern Physics*, 47:55–61.

de Broglie, L. (1928). La nouvelle dynamique des quanta. *Electrons et photons. Rapports et discussions du cinquième Conseil de physique tenu à Bruxelles du 24 au 29 octobre 1927 sous les auspices de l'Institut international de physique Solvay*, pages 105–132. Paris: Gauthier-Villars. English translation in Bacciagaluppi, G. and Valentini, A. (2009). *Quantum theory at the crossroads. Reconsidering the 1927 Solvay conference*, pages 341–371. Cambridge: Cambridge University Press.

de Broglie, L. (1964). *The current interpretation of wave mechanics. A critical study*. Amsterdam: Elsevier.

Deckert, D.-A., Dürr, D., Merkl, F., and Schottenloher, M. (2010). Time-evolution of the external field problem in quantum electrodynamics. *Journal of Mathematical Physics*, 51(12):122301.

Deckert, D.-A., Dürr, D., and Vona, N. (2014). Delay equations of the Wheeler-Feynman type. *Journal of Mathematical Sciences*, 202(5):623–636.

Deckert, D.-A. and Hartenstein, V. (2016). On the initial value formulation of classical electrodynamics. *Preprint: arXiv:1602.0468*.

Deckert, D.-A. and Hinrichs, G. (2016). Electrodynamic two-body problem for prescribed initial data on a straight line. *Journal of Differential Equations*, 260 (9):6900–6929.

Dennett, D. C. (1988). Quining qualia. In Marcel, A. and Bisiach, E., editors, *Consciousness in contemporary science*, pages 43–77. Oxford: Oxford University Press.

Dickson, M. (2000). Are there material objects in Bohm's theory? *Philosophy of Science*, 67(4):704–710.

Dieks, D. (2001). Space-time relationalism in Newtonian and relativistic physics. *International Studies in the Philosophy of Science*, 15(1):5–17.

Dieks, D. (2006). Becoming, relativity and locality. In Dieks, D., editor, *The ontology of space-time*, pages 157–176. Amsterdam: Elsevier.

Dieks, D. and Versteegh, M. A. M. (2008). Identical quantum particles and weak discernability. *Foundations of Physics*, 38:923–934.

Dirac, P. A. M. (1934). Théorie du positron. In *Rapport du 7ème Conseil Solvay de physique: structure et propriétés des noyaux atomiques*, pages 203–212. Paris: Gauthier-Villars. Reprinted in Julian Schwinger (ed.): *Selected papers on quantum electrodynamics*. New York: Dover 1958. Chapter 7, pp. 82–91.

Dirac, P. A. M. (1938). Classical theory of radiating electrons. *Proceedings of the Royal Society A*, 167(929):148–169.

Dirac, P. A. M. (1947). *The principles of quantum mechanics. Third edition*. Oxford: Oxford University Press.

Dizadji-Bahmani, F. (2015). The probability problem in Everettian quantum mechanics persists. *British Journal for the Philosophy of Science*, 66:257–283.

Dowe, P. (1996). Backwards causation and the direction of causal processes. *Mind*, 105:227–248.

Dowker, F. and Herbauts, I. (2005). The status of the wave function in dynamical collapse models. *Foundations of Physics Letters*, 18:499–518.

Dürr, D., Goldstein, S., Norsen, T., Struyve, W., and Zanghì, N. (2013a). Can Bohmian mechanics be made relativistic? *Proceedings of the Royal Society A*, 470:2162.

Dürr, D., Goldstein, S., Taylor, J., Tumulka, R., and Zanghì, N. (2006). Topological factors derived from Bohmian mechanics. *Annales Henri Poincaré*, 7(4):791–807. Reprinted in Dürr et al. (2013b, ch. 8).

Dürr, D., Goldstein, S., Tumulka, R., and Zanghì, N. (2005). Bell-type quantum field theories. *Journal of Physics A: Mathematical and General*, 38(4):R1–R43.

Dürr, D., Goldstein, S., and Zanghì, N. (2013b). *Quantum physics without quantum philosophy*. Berlin: Springer.

Dürr, D. and Teufel, S. (2009). *Bohmian mechanics: the physics and mathematics of quantum theory*. Berlin: Springer.

Earman, J. (1989). *World enough and space-time. Absolute versus relational theories of spacetime*. Cambridge, Massachusetts: MIT Press.

Earman, J. (2011). Sharpening the electromagnetic arrow(s) of time. In Callender, C., editor, *The Oxford handbook of philosophy of time*, chapter 16, pages 360–391. Oxford: Oxford University Press.

Earman, J. and Norton, J. (1987). What price spacetime substantivalism? The hole story. *British Journal for the Philosophy of Science*, 38:515–525.

Egg, M. and Esfeld, M. (2014). Non-local common cause explanations for EPR. *European Journal for Philosophy of Science*, 4:181–196.

Egg, M. and Esfeld, M. (2015). Primitive ontology and quantum state in the GRW matter density theory. *Synthese*, 192(10):3229–3245.

Einstein, A. (1948). Quanten-Mechanik und Wirklichkeit. *Dialectica*, 2:320–324.

Einstein, A. and Infeld, L. (1949). On the motion of particles in general relativity theory. *Canadian Journal of Mathematics*, 1:209–241.

Einstein, A., Podolsky, B., and Rosen, N. (1935). Can quantum-mechanical description of physical reality be considered complete? *Philosophical Review*, 47:777–780.

Esfeld, M. (2001). *Holism in philosophy of mind and philosophy of physics*. Dordrecht: Kluwer.

Esfeld, M. (2004). Quantum entanglement and a metaphysics of relations. *Studies in History and Philosophy of Modern Physics*, 35:601–617.

Esfeld, M. (2009). The modal nature of structures in ontic structural realism. *International Studies in the Philosophy of Science*, 23:179–194.

Esfeld, M. (2014a). The primitive ontology of quantum physics: guidelines for an assessment of the proposals. *Studies in History and Philosophy of Modern Physics*, 47:99–106.

Esfeld, M. (2014b). Quantum Humeanism, or: physicalism without properties. *Philosophical Quarterly*, 64(256):453–470.

Esfeld, M. (2015). Bell's theorem and the issue of determinism and indeterminism. *Foundations of Physics*, 45:471–482.

Esfeld, M. and Gisin, N. (2014). The GRW flash theory: a relativistic quantum ontology of matter in space-time? *Philosophy of Science*, 81:248–264.

Esfeld, M. and Lam, V. (2008). Moderate structural realism about space-time. *Synthese*, 160(1):27–46.

Esfeld, M. and Lam, V. (2011). Ontic structural realism as a metaphysics of objects. In A. and Bokulich, P., editors, *Scientific structuralism*, pages 143–159. Dordrecht: Springer.

Esfeld, M. and Lam, V. (2013). A dilemma for the emergence of spacetime in canonical quantum gravity. *Studies in History and Philosophy of Modern Physics*, 44:286–293.

Esfeld, M., Lazarovici, D., Hubert, M., and Dürr, D. (2014). The ontology of Bohmian mechanics. *British Journal for the Philosophy of Science*, 65(4):773–796.

Everett, H. (1957). "Relative state" formulation of quantum mechanics. *Reviews of Modern Physics*, 29(3):454–462. Reprinted in DeWitt, B. S. and Graham, N., editors (1973). *The Many-Worlds Interpretation of Quantum Mechanics*, pages 141–149. Princeton: Princeton University Press.

Feynman, R. P. (1966). The development of the space-time view of quantum electrodynamics. Nobel Lecture, December 11, 1965. *Science*, 153:699–708.

Feynman, R. P., Leighton, R. B., and Sands, M. (1963). *The Feynman lectures on physics. Volume 1*. Reading (Massachusetts): Addison-Wesley.

Field, H. H. (1980). *Science without numbers. A defence of nominalism*. Oxford: Blackwell.

Field, H. H. (1985). Can we dispense with space-time? In Asquith, P. D. and Kitcher, P., editors, *Proceedings of the 1984 Biennial Meeting of the Philosophy of Science Association. Volume 2*, pages 33–90. East Lansing: Philosophy of Science Association.

Fierz, H. and Scharf, G. (1979). Particle interpretation for external field problems in QED. *Helvetica Physica Acta. Physica Theoretica*, 52(4):437–453.

Floridi, L. (2008). A defence of informational structural realism. *Synthese*, 161:219–253.

Fokker, A. D. (1929). Ein invarianter Variationssatz für die Bewegung mehrerer elektrischer Massenteilchen. *Zeitschrift für Physik*, 58(5):386–393.

Foster, J. (1982). *The case for idealism*. London: Routledge.

Frankel, T. (1997). *The geometry of physics*. Cambridge: Cambridge University Press.

French, S. (2010). The interdependence of objects, structure and dependence. *Synthese*, 175:89–109.

French, S. (2014). *The structure of the world. Metaphysics and representation.* Oxford: Oxford University Press.

Frisch, M. (2005). *Inconsistency, asymmetry, and non-locality. A philosophical investigation of classical electrodynamics.* New York: Oxford University Press.

Gauss, C. F. (1867). Brief an Wilhelm Weber 19. März 1845. In *Werke. Mathematische Physik. Band 5*, pages 627–629. Göttingen: Universitätsdruckerei.

Geach, P. (1965). Some problems about time. *Proceedings of the British Academy*, 51:321–336.

Gerhardt, C. I., editor (1890). *Die philosophischen Schriften von G. W. Leibniz. Band 7*. Berlin: Weidmannsche Verlagsbuchhandlung.

Ghirardi, G. C., Grassi, R., and Benatti, F. (1995). Describing the macroscopic world: closing the circle within the dynamical reduction program. *Foundations of Physics*, 25(1):5–38.

Ghirardi, G. C., Pearle, P., and Rimini, A. (1990). Markov processes in Hilbert space and continuous spontaneous localization of systems of identical particles. *Physical Review A*, 42:78–89.

Ghirardi, G. C., Rimini, A., and Weber, T. (1986). Unified dynamics for microscopic and macroscopic systems. *Physical Review D*, 34(2):470–491.

Gisin, N. (1984). Quantum measurements and stochastic processes. *Physical Review Letters*, 52:1657–1660 (and reply p. 1776).

Gisin, N. (1989). Stochastic quantum dynamics and relativity. *Helvetica Physica Acta*, 62:363–371.

Goldstein, S. (2012). *Probability in physics*, chapter Typicality and notions of probability in physics, pages 59–71. Berlin: Springer.

Goldstein, S. and Struyve, W. (2007). On the uniqueness of quantum equilibrium in Bohmian mechanics. *Journal of Statistical Physics*, 128(5):1197–1209.

Goldstein, S., Taylor, J., Tumulka, R., and Zanghì, N. (2005a). Are all particles identical? *Journal of Physics A: Mathematical and General*, 38(7):1567–1576.

Goldstein, S., Taylor, J., Tumulka, R., and Zanghì, N. (2005b). Are all particles real? *Studies in History and Philosophy of Modern Physics*, 36(1):103–112.

Gomes, H., Gryb, S., and Koslowski, T. (2011). Einstein gravity as a 3D conformally invariant theory. *Classical and Quantum Gravity*, 28:045005.

Gomes, H. and Koslowski, T. (2012). Coupling shape dynamics to matter gives spacetime. *General relativity and gravitation*, 44(6):1539–1553.

Gomes, H. and Koslowski, T. (2013). Frequently asked questions about shape dynamics. *Foundations of Physics*, 43(12):1428–1458.

Graham, D. W. (2010). *The texts of early Greek philosophy. The complete fragments and selected testimonies of the major Presocratics. Edited and translated by Daniel W. Graham.* Cambridge: Cambridge University Press.

Gravejat, P., Hainzl, C., Lewin, M., and Séré, E. (2013). Construction of the Pauli–Villars-regulated Dirac vacuum in electromagnetic fields. *Archive for Rational Mechanics and Analysis*, 208(2):603–665.

Greiner, W. and Bromley, D. A. (2000). *Relativistic quantum mechanics. Third edition. Wave equations.* Berlin: Springer.

Gryb, S. and Thébault, K. P. Y. (2016). Time remains. *British Journal for the Philosophy of Science*, 67:663–705.

Guay, A. and Pradeu, T. (2017). Right out of the box: How to situate metaphysics of science with regard to other metaphysical approaches. *Synthese*, forthcoming.

Hacking, I. (1975). The identity of indiscernibles. *Journal of Philosophy*, 72:249–256.

Hall, M. J. W., Deckert, D.-A., and Wiseman, H. M. (2014). Quantum phenomena modelled by interactions between many classical worlds. *Physical Review X*, 4:041013.

Hall, N. (2009). Humean reductionism about laws of nature. Unpublished manuscript. http://philpapers.org/rec/halhra.

Halvorson, H. and Clifton, R. K. (2002). No place for particles in relativistic quantum theories? *Philosophy of Science*, 69:1–28.

Hoffmann-Kolss, V. (2010). *The metaphysics of extrinsic properties*. Frankfurt (Main): Ontos.

Holland, P. (2001a). Hamiltonian theory of wave and particle in quantum mechanics I: Liouville's theorem and the interpretation of the the de Broglie-Bohm theory. *Il Nuovo Cimento B*, 116:1043–1070.

Holland, P. (2001b). Hamiltonian theory of wave and particle in quantum mechanics II: Hamilton-Jacobi theory and particle back-reaction. *Il Nuovo Cimento B*, 116:1143–1172.

Huggett, N. (2006). The regularity account of relational spacetime. *Mind*, 115 (457):41–73.

Huggett, N. and Wüthrich, C. (2013). Emergent spacetime and empirical (in)coherence. *Studies in History and Philosophy of Modern Physics*, 44:276–285.

Hume, D. (1739). *A treatise of human nature*.

Hume, D. (1748). *Enquiries concerning human understanding and concerning the principles of morals*.

Humphreys, P. (2013). Scientific ontology and speculative ontology. In Ross, D., Ladyman, J., and Kincaid, H., editors, *Scientific metaphysics*, pages 51–78. Oxford: Oxford University Press.

Jackson, F. (1994). Armchair metaphysics. In Michael, M. and Hawthorne, J. O., editors, *Philosophy in mind. The place of philosophy in the study of mind*, pages 23–42. Dordrecht: Kluwer.

Jackson, F. (1998). *From metaphysics to ethics. A defence of conceptual analysis*. Oxford: Oxford University Press.

Keränen, J. (2001). The identity problem for realist structuralism. *Philosophia Mathematica*, 9(3):308–330.

Kitcher, P. (1981). Explanatory unification. *Philosophy of Science*, 48:501–531.

Klein, O. (1929). Die Reflexion von Elektronen an einem Potentialsprung nach der relativistischen Dynamik von Dirac. *Zeitschrift für Physik*, 53(3-4):157–165.

Kuhlmann, M. (2010). *The ultimate constituents of the material world. In search of an ontology for fundamental physics*. Frankfurt (Main): Ontos.

Ladyman, J. (2007). On the identity and diversity of objects in a structure. *Proceedings of the Aristotelian Society. Supplementary Volume*, 81(1):23–43.

Ladyman, J. and Bigaj, T. (2010). The principle of the identity of indiscernibles and quantum mechanics. *Philosophy of Science*, 77:117–136.

Ladyman, J. and Ross, D. (2007). *Every thing must go: metaphysics naturalized*. New York: Oxford University Press.

Lanczos, C. (1970). *The variational principles of mechanics. Fourth edition*. University of Toronto Press.

Lange, M. (2002). *An introduction to the philosophy of physics: locality, fields, energy and mass*. Oxford: Blackwell.

Lange, M. (2013). Grounding, scientific explanation, and Humean laws. *Philosophical Studies*, 164:255–261.

Langton, R. and Lewis, D. (1998). Defining 'intrinsic'. *Philosophy and Phenomenological Research*, 58:333–345.

Laudisa, F. (2014). Against the 'no-go' philosophy of quantum mechanics. *European Journal for Philosophy of Science*, 4:1–17.

Lazarovici, D. (2014). A relativistic retrocausal model violating Bell's inequality. *Proceedings of the Royal Society A*, 471:DOI 10.1098/rspa.2014.0454.

Lazarovici, D. and Reichert, P. (2015). Typicality, irreversibility and the status of macroscopic laws. *Erkenntnis*, 80(4):689–716.

LeBihan, B. (2016). Super-relationalism: combining eliminativism about objects and relationism about spacetime. *Philosophical Studies*, 173:2151–2172.

Lehmkuhl, D. (2016). The metaphysics of super-substantivalism. *Noûs*, DOI 10.1111/nous.12163.

Lewis, D. (1973a). Causation. *Journal of Philosophy*, 70:556–567.

Lewis, D. (1973b). *Counterfactuals*. Oxford: Blackwell.

Lewis, D. (1986a). *On the plurality of worlds*. Oxford: Blackwell.

Lewis, D. (1986b). *Philosophical papers*, volume 2. Oxford: Oxford University Press.

Lewis, D. (1994). Humean supervenience debugged. *Mind*, 103(412):473–490. Reprinted in Lewis, D. (1999). *Papers in Metaphysics and Epistemology*, pages 224–247. Cambridge: Cambridge University Press.

Lewis, D. (2004). Causation as influence. In Collins, J., Hall, N., and Paul, L. A., editors, *Causation and counterfactuals*, pages 75–106. Cambridge, Massachusetts: MIT Press.

Lewis, D. (2009). Ramseyan humility. In Braddon-Mitchell, D. and Nola, R., editors, *Conceptual Analysis and Philosophical Naturalism*, pages 203–222. Cambridge, Massachusetts: MIT Press.

Lewis, P. J. (1997). Quantum mechanics, orthogonality, and counting. *British Journal for the Philosophy of Science*, 48:313–328.

Locke, J. (1690). *An essay concerning human understanding*.

Loewer, B. (1996). Freedom from physics: quantum mechanics and free will. *Philosophical Topics*, 24(2):91–112.

Loewer, B. (2007). Counterfactuals and the second law. In Price, H. and Corry, R., editors, *Causation, physics, and the constitution of reality. Russell's republic revisited*, pages 293–326. Oxford: Oxford University Press.

Loewer, B. (2012). Two accounts of laws and time. *Philosophical Studies*, 160:115–137.

Mach, E. (1919). *The science of mechanics: a critical and historical account of its development. Fourth edition. Translation by Thomas J. McCormack*. Chicago: Open Court.

Malament, D. B. (1977). The class of continuous timelike curves determines the topology of spacetime. *Journal of Mathematical Physics*, 18(7):1399–1404.

Marshall, D. (2015). Humean laws and explanation. *Philosophical Studies*, 172:3145–3165.

Martens, N. (2017). Regularity comparativism about mass in Newtonian gravity. *Philosophy of Science*, 84(5).

Maudlin, T. (2007a). *The metaphysics within physics*. New York: Oxford University Press.

Maudlin, T. (2007b). What could be objective about probabilities? *Studies in History and Philosophy of Modern Physics*, 38:275–291.

Maudlin, T. (2010). Can the world be only wave-function? In Saunders, S., Barrett, J., Kent, A., and Wallace, D., editors, *Many worlds? Everett, quantum theory, and reality*, pages 121–143. Oxford: Oxford University Press.

Maudlin, T. (2011). *Quantum non-locality and relativity. Third edition.* Chichester: Wiley-Blackwell.

Maudlin, T. (2014). Critical study of David Wallace, The emergent multiverse: quantum theory according to the Everett interpretation, Oxford University Press 2012. *Noûs*, 48(4):794–808.

Maudlin, T. (2015). The universal and the local in quantum theory. *Topoi*, 34:349–358.

McKenzie, K. (2014). Priority and particle physics: ontic structural realism as a fundamentality thesis. *British Journal for the Philosophy of Science*, 65:353–380.

Mickelsson, J. (2014). The phase of the scattering operator from the geometry of certain infinite-dimensional groups. *arXiv:1403.4340 [math-ph]*.

Miller, E. (2014). Quantum entanglement, Bohmian mechanics, and Humean supervenience. *Australasian Journal of Philosophy*, 92:567–583.

Miller, E. (2015). Humean scientific explanation. *Philosophical Studies*, 172:1311–1332.

Misner, C. W., Thorne, K. S., and Wheeler, J. A. (1973). *Gravitation.* San Francisco: Freeman.

Monton, B. (2004). The problem of ontology for spontaneous collapse theories. *Studies in History and Philosophy of Modern Physics*, 35(3):407–421.

Morganti, M. (2013). *Combining science and metaphysics. Contemporary physics, conceptual revision and common cause.* Basingstoke: Palgrave Macmillan.

Muller, F. A. (2011). How to defeat Wüthrich's abysmal embarassment argument against space-time structuralism. *Philosophy of Science*, 78:1046–1057.

Mulligan, K. (2012). Beyond objects, properties and relations. Handout of a lecture given at the University of Lausanne.

Mundy, B. (1989). Distant action in classical electromagnetic theory. *British Journal for the Philosophy of Science*, 40(1):39–68.

Newton, I. (1952). *Opticks. Edited by I. B. Cohen.* New York: Dover.

Ney, A. (2012). Neo-positivist metaphysics. *Philosophical Studies*, 160:53–78.

Ney, A. (2015). Fundamental physical ontologies and the constraint of empirical coherence: a defense of wave function realism. *Synthese*, 192(10):3105–3124.

Norsen, T. (2005). Einstein's boxes. *American Journal of Physics*, 73:164–176.

Pickl, P. (2011). A simple derivation of mean field limits for quantum systems. *Letters in Mathematical Physics*, 97(2):151–164.

Pooley, O. (2001). Relationalism rehabilitated? II: Relativity. http://philsci-archive.pitt.edu/221/.

Pooley, O. (2013). Substantivalist and relationalist approaches to spacetime. In Batterman, R., editor, *The Oxford handbook of philosophy of physics*, pages 522–586. Oxford: Oxford University Press.

Pooley, O. and Brown, H. (2002). Relationalism rehabilitated? I: Classical mechanics. *British Journal for the Philosophy of Science*, 53:183–204.

Price, H. (1996). *Time's arrow and Archimedes' point. New directions for the physics of time.* Oxford: Oxford University Press.

Pylkkänen, P., Hiley, B. J., and Pättiniemi, I. (2015). Bohm's approach and individuality. In Guay, A. and Pradeu, T., editors, *Individuals across the sciences*, chapter 12, pages 226–246. Oxford: Oxford University Press.

Quine, W. V. O. (1948). On what there is. *Review of Metaphysics*, 2:21–38.

Quine, W. V. O. (1969). *Ontological relativity and other essays.* New York: Coulmbia University Press.

Ritz, W. (1908). Recherches critiques sur l'électrodynamique générale. *Annales de chimie et de physique*, 8(13):145–209.

Robinson, H. (1982). *Matter and sense. A critique of contemporary materialism.* Cambridge: Cambridge University Press.

Rohrlich, F. (1997). The dynamics of a charged sphere and the electron. *American Journal of Physics*, 65(11).

Ross, D., Ladyman, J., and Kincaid, H., editors (2013). *Scientific metaphysics*. Oxford: Oxford University Press.

Rovelli, C. (1997). Halfway through the woods: contemporary research on space and time. In Earman, J. and Norton, J., editors, *The cosmos of science*, pages 180–223. Pittsburgh: University of Pittsburgh Press.

Ruetsche, L. (2011). *Interpreting quantum theories. The art of the possible*. Oxford: Oxford University Press.

Ruijsenaars, S. N. M. (1977). Charged particles in external fields. I. Classical theory. *Journal of Mathematical Physics*, 18(4):720–737.

Russell, B. (1912). On the notion of cause. *Proceedings of the Aristotelian Society*, 13:1–26.

Russell, B. (1927). *The analysis of matter*. London: Routledge.

Saunders, S. (1991). The negative-energy sea. In Saunders, S. and Brown, H. R., editors, *The philosophy of vacuum*, pages 65–109. Oxford: Oxford University Press.

Saunders, S. (1999). The 'beables' of relativistic pilot-wave theory. In Butterfield, J. and Pagonis, C., editors, *From physics to philosophy*, pages 70–89. Cambridge: Cambridge University Press.

Saunders, S. (2006). Are quantum particles objects? *Analysis*, 66:52–63.

Saunders, S. (2013). Rethinking Newton's Principia. *Philosophy of Science*, 80 (1):22–48.

Schaffer, J. (2010a). The internal relatedness of all things. *Mind*, 119(474):341–376.

Schaffer, J. (2010b). Monism: the priority of the whole. *Philosophical Review*, 119 (1):31–76.

Scharf, G. (1995). *Finite quantum electrodynamics: the causal approach. Second edition*. Berlin: Springer.

Schiff, J. and Poirier, B. (2012). Communication: quantum mechanics without wavefunctions. *Journal of Chemical Physics*, 136(3):031102.

Schrödinger, E. (1930). Über die kräftefreie Bewegung in der relativistischen Quantenmechanik. *Sitzungsberichte der Preußischen Akademie der Wissenschaften. Physikalisch-Mathematische Klasse*, pages 418–428.

Schwarzschild, K. (1903). Zur Elektrodynamik. ii. Die elementare elektrodynamische Kraft. *Nachrichten von der Gesellschaft der Wissenschaften zu Göttingen, Mathematisch-Physikalische Klasse*, 1903:132–141.

Sebens, C. (2015). Quantum mechanics as classical physics. *Philosophy of Science*, 82:266–291.

Seevinck, M. P. and Uffink, J. (2011). Not throwing out the baby with the bathwater: Bell's condition of local causality mathematically 'sharp and clean'. In Dieks, D., Gonzalez, W. J., Hartmann, S., Uebel, T., and Weber, M., editors, *Explanation, prediction and confirmation*, pages 425–450. Dordrecht: Springer.

Sellars, W. (1956). Empiricism and the philosophy of mind. In Feigl, H. and Scriven, M., editors, *The foundations of science and the concepts of psychology and psychoanalysis. Minnesota Studies in the Philosophy of Science. Volume 1*, pages 253–329. Minneapolis: University of Minnesota Press.

Sellars, W. (1962). Philosophy and the scientific image of man. In Colodny, R., editor, *Frontiers of science and philosophy*, pages 35–78. Pittsburgh: University of Pittsburgh Press.

Sklar, L. (1974). *Space, time and spacetime*. Berkeley: University of California Press.

Skow, B. (2015). *Objective becoming*. Oxford: Oxford University Press.

Spohn, H. (2000). The critical manifold of the Lorentz-Dirac equation. *EPL (Europhysics Letters)*, 50(3):287.

Strawson, G. (1989). *The secret connexion. Causation, realism, and David Hume.* Oxford: Oxford University Press.

Struyve, W. (2010). Pilot-wave approaches to quantum field theory. *Journal of Physics: Conference Series*, 306:012047.

Struyve, W. and Westman, H. (2006). A new pilot-wave model for quantum field theory. *AIP Conference Proceedings*, 844:321–339.

Struyve, W. and Westman, H. (2007). A minimalist pilot-wave model for quantum electrodynamics. *Proceedings of the Royal Society A*, 463:3115–3129.

Suárez, M. (2015). Bohmian dispositions. *Synthese*, 192:3203–3228.

Sutherland, R. I. (2008). Causally symmetric Bohm model. *Studies in History and Philosophy of Modern Physics*, 39:782–805.

Tetrode, H. (1922). Über den Wirkungszusammenhang der Welt. Eine Erweiterung der klassischen Dynamik. *Zeitschrift für Physik*, 10(1):317–328.

Tumulka, R. (2006). A relativistic version of the Ghirardi-Rimini-Weber model. *Journal of Statistical Physics*, 125(4):821–840.

Tumulka, R. (2009). The point processes of the GRW theory of wave function collapse. *Reviews in Mathematical Physics*, 21:155–227.

Tumulka, R. (2011). Paradoxes and primitive ontology in collapse theories of quantum mechanics. *arXiv:1102.5767 [quant-ph]*.

Unruh, W. G. (1976). Notes on black-hole evaporation. *Phys. Rev. D*, 14:870–892.

Valentini, A. (1992). *On the pilot-wave theory of classical, quantum and subquantum physics.* Trieste: International School for Advanced Studies, PhD thesis.

Vassallo, A. (2015). Can Bohmian mechanics be made background independent? *Studies in History and Philosophy of Modern Physics*, 52:242–250.

Vassallo, A. and Ip, P. H. (2016). On the conceptual issues surrounding the notion of relational Bohmian dynamics. *Foundations of Physics*, 46(8):943–972.

Wallace, D. (2008). Philosophy of quantum mechanics. In Rickles, D., editor, *The Ashgate companion to contemporary philosophy of physics*, pages 16–98. Aldershot: Ashgate.

Wallace, D. (2012). *The emergent multiverse. Quantum theory according to the Everett interpretation.* Oxford: Oxford University Press.

Wallace, D. (2014). Life and death in the tails of the GRW wave function. *arXiv:1407.4746 [quant-ph]*.

Weyl, H. (1949). *Philosophy of mathematics and natural science.* Princeton: Princeton University Press.

Wheeler, J. A. (1962). *Geometrodynamics.* New York: Academic Press.

Wheeler, J. A. and Feynman, R. P. (1945). Interaction with the absorber as the mechanism of radiation. *Reviews of Modern Physics*, 17:157–181.

Wilson, A. (2013a). Objective probability in Everettian quantum mechanics. *British Journal for the Philosophy of Science*, 64:709–737.

Wilson, M. (2013b). What can contemporary philosophy learn from our 'scientific philosophy' heritage? In Ross, D., Ladyman, J., and Kincaid, H., editors, *Scientific metaphysics*, pages 151–181. Oxford: Oxford University Press.

Wüthrich, C. (2009). Challenging the spacetime structuralist. *Philosophy of Science*, 76:1039–1051.

Zeh, H. D. (2010). *The physical basis of the direction of time. Fifth edition.* Berlin: Springer.

Index

3 20